HeadStart ✓
Primary

MASTERING THE MATHEMATICS CURRICULUM
YEAR 1

Written by Laura Sumner

Acknowledgements:

Author: Laura Sumner
Cover and Page Design: Jerry Fowler
Illustration: Jerry Fowler and Kathryn Webster

The right of Laura Sumner to be identified as the author of this publication has been asserted by her in accordance with the Copyright, Designs and Patents Act 1998.

HeadStart Primary Ltd
Elker Lane
Clitheroe
BB7 9HZ

T. 01200 423405
E. info@headstartprimary.com
www.headstartprimary.com

Published by HeadStart Primary Ltd 2017 © **HeadStart Primary Ltd 2017 (V3)**

A record for this book is available from the British Library
ISBN: 978-1-908767-51-6

MASTERING THE MATHEMATICS CURRICULUM
Teachers' Notes – Year 1

RATIONALE

Background

This book has been written taking into account the principles outlined in the current Mathematics curriculum. It focuses on a mastery approach to teaching mathematics, as outlined by NCETM's director, Charlie Stripp, in the short paper entitled 'Mastery approaches to mathematics and the new national curriculum'. In addition, account is taken of the Oxford University Press series 'Teaching for Mastery.'

Achieving mastery

Emphasis is placed on the importance of children fully mastering mathematical concepts and principles so that, as well as being able to complete mathematical tasks, they are able to gain a deep understanding about why mathematical procedures work. Underlying this, is an expectation that all children can accomplish high standards. Consequently, teachers need to be able to focus their time on planning highly effective strategies to teach and model mathematical concepts for children.

Charlie Stripp highlights the value of well-planned and focused lessons, where teachers explore and enhance children's understanding through highly effective and precise questioning. Additionally, he points to the importance of effective practice and consolidation of key concepts.

Therefore, the over-riding aim of this book is to provide examples for children to practise the concepts they are being taught, thereby allowing teachers more time to focus on achieving high quality teaching. Consequently, for each objective within the mathematics curriculum, as outlined in the 'Statutory requirements' and the 'Notes and guidance', there is at least one page of pertinent questions which children can complete during a lesson or for homework. For objectives such as 'performing mental calculations', a number of pages are available for children to practise a range of appropriate strategies. Focused discussion, with teachers and other adults, about the concepts highlighted will enhance children's depth of understanding and mastery of the curriculum.

Embedding mastery

Opportunities for children to practise and demonstrate their mastery of mathematical concepts run through all the pages within the book. So that children can gain an in-depth understanding of each concept, the book is designed to provide examples precisely matched to each specific objective. Having this deep understanding will enable children to apply their knowledge and skills to different contexts, sometimes requiring an understanding of a range of objectives and concepts. Therefore, the book also includes further pages to provide an opportunity for children to practise and demonstrate this level of mastery. So, for each domain within the curriculum, there are specific pages linked to each objective, as well as further mastery pages which draw together the concepts within the whole domain.

DIFFERENTIATION

The 'Teaching for Mastery' Oxford University Press series recommends that the class should work "together on the same topic, whilst at the same time addressing the need for all pupils to master the curriculum and for some to gain greater depth of proficiency and understanding". The NCETM paper 'Mastery approaches to teaching mathematics and the new national curriculum' suggests that "differentiation occurs in the support and intervention provided to different pupils".

The book is designed to meet these recommendations. Some groups of children may benefit from additional adult support and intervention to complete the examples, and develop their proficiency and understanding. In general, the examples are arranged so that they become progressively more difficult within each question or on each page. The 'Teaching for Mastery' series suggests that there should not be a need for additional 'catch-up programmes'. However, some groups may require targeted intervention to reinforce their understanding of their own year group's concepts through carrying out work at a level appropriate to their current ability. Therefore, some children could work on relevant pages from the HeadStart books for other year groups, whilst still working on the same topic as the rest of the class.

ASSESSING CHILDREN'S PROGRESS

This book is not designed to be used as a summative tool. Nevertheless, as the book is based on the Year 1 expectations within the mathematics curriculum, it can support teachers in making formative assessments about children's progress towards those expectations. Furthermore, the book can provide diagnostic information about aspects of the curriculum for which individual children, or groups of children, may need further support or enhancement.

USING THE BOOK

For younger children, who are still developing reading skills, it may be beneficial for some pages to be read by an adult so that children are able to focus on the mathematical content and concepts.

The book is designed so that children are able to write answers on the photocopied or printed sheets, which may be particularly useful if given as homework. However, for most pages, pupils can easily transcribe the work into their exercise books. If the children complete work on photocopied or printed sheets and substantial 'working out' is to be completed, this may need to be carried out separately or in exercise books.

In addition to photocopying or printing the pages for the children's use, the enclosed CD can be used to project pages onto an interactive whiteboard, thereby enabling them to be used for modelling and clarification purposes.

CONTENTS

Number – number and place value

Number – addition and subtraction

CONTENTS

Number – multiplication and division

Number – fractions

CONTENTS

CONTENTS

FURTHER MASTERY PAGES FOR EACH DOMAIN

HeadStart Primary

Number

Number and place value

YEAR 1

Name .... Class .... Date ....

Count forwards, read and write numbers to and across 100, beginning with 0 or 1, or from any given number

1 Start at **0** and count to **100**.

2 Now, start at **52** and count to **108**.

3 Try to write the missing numbers on the animals.

a

b

c

d

e

1

Number – number and place value

YEAR 1

Name .... Class .... Date ....

Count backwards, read and write numbers from any given number up to and beyond 100

1 Start at **10** and count backwards to **0**.

2 Now, start at **107** and count backwards to **84**.

3 Try to write the missing numbers on the fruit.

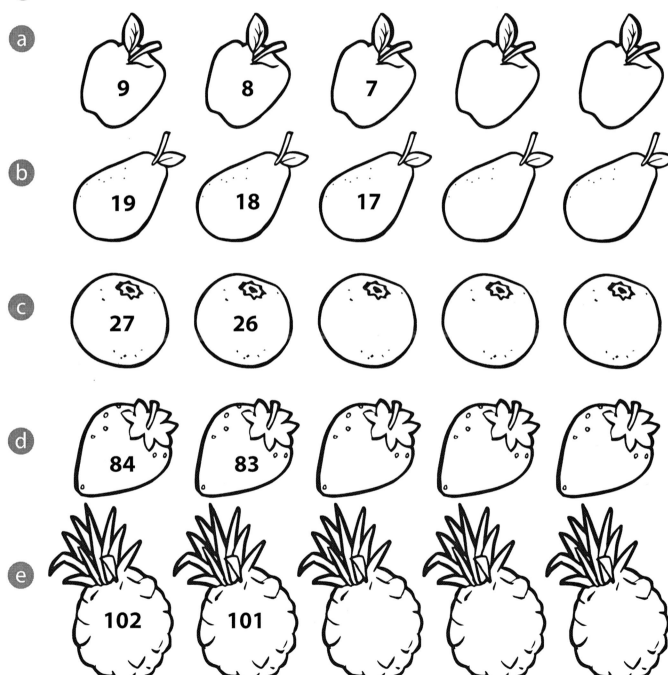

a   9   8   7

b   19   18   17

c   27   26

d   84   83

e   102   101

Name .... Class .... Date ....

Use ordinal numbers

1 The children are lining up at school. Match the ordinal numbers. One is done for you.

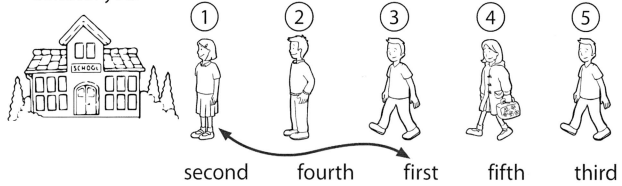

second   fourth   first   fifth   third

2 Write a word to complete the sentences. One is done for you.

The | first | toy is the doll  .

a The | | toy is the teddy .

b The | | toy is the train .

3 Write the ordinal numbers on the stepping stones. Some are done for you.

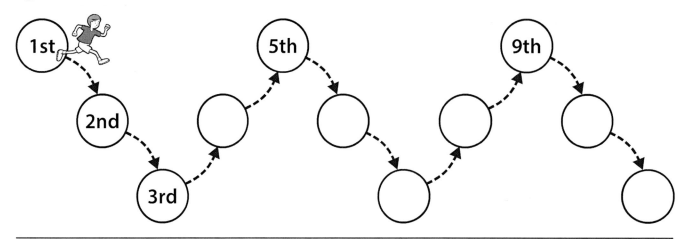

Name .... Class .... Date ....

Count, read and write numbers to 100 to indicate a quantity

1 How many sweets can you see? ☐

2 How many carrots are there? ☐

3 How many lollipops are there? ☐

4 How long is the pencil? | centimetres |

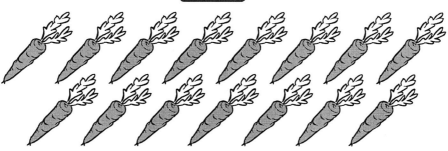

not actual size

0   cm   30

5 Count on from the box of pencils to find how many pencils there are altogether. ☐

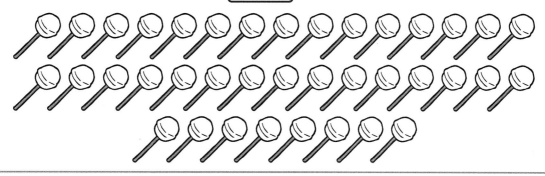

PENCILS
50

Name .... Class .... Date ....

Count in twos

1 Count in **twos** to find the answer.

a How many socks can you see? ☐

b How many shoes are there altogether? ☐

c How many ties are there in total? ☐

2 Now count in **twos** and write the missing numbers on the shapes.

a 2 4 ☆ ☆ ☆ ☆

b 10 △ 14 △ △ △

c 28 ◯ ◯ 34 ◯ ◯

d 17 19 ☐ ☐ ☐ ☐

Name .... Class .... Date ....

Count in fives

1 Count in **fives** to find the answers.

a How many balls are there altogether? [ ]

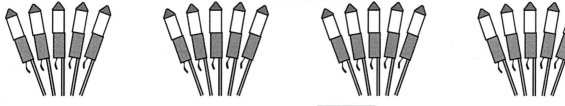

b How many rockets can you see? [ ]

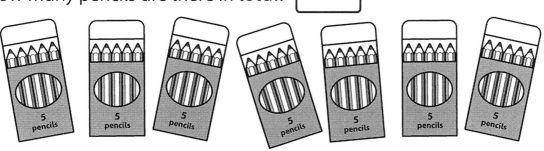

c How many pencils are there in total? [ ]

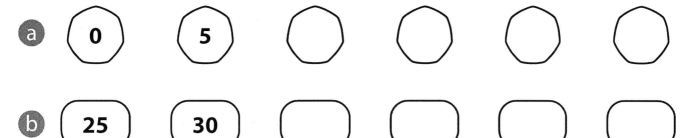

2 Now count in **fives** and write the missing numbers on the shapes.

a (0) (5) ( ) ( ) ( ) ( )

b (25) (30) ( ) ( ) ( ) ( )

c (60) (65) ( ) ( ) ( ) ( )

d (28) (33) ( ) ( ) ( ) ( )

Name .... Class .... Date ....

Count in tens

1 Count in **tens** to find the answer.

a How many sweets can you see? ☐

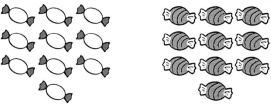

b How many mushrooms are there? ☐

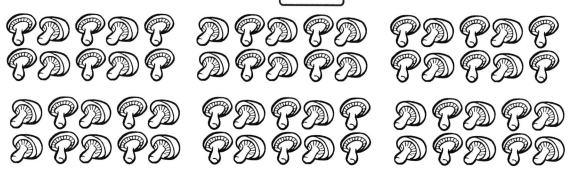

c How many biscuits are there? ☐

- -

2 Now count in **10s** and write the missing numbers in the boxes.

a | 0 | 10 | ⬭ | ⬭ | ⬭ | ⬭ |

b | 40 | 50 | ⬭ | ⬭ | ⬭ | ⬭ |

c | 65 | ⬭ | 85 | ⬭ | ⬭ | ⬭ |

d | 33 | ⬭ | 53 | ⬭ | ⬭ | ⬭ |

Name .... Class .... Date ....

Recognise odd and even numbers when counting

1 Draw rings around the <u>odd</u> numbers.

12   17   8   4   3   9

2 Count from **0** to **20** and write down all the <u>even</u> numbers.

0
....
                                              20
....

3 Now count from **20** to **40** and write down all the <u>odd</u> numbers.

....

....

4 Start from **5** and count in **fives** to **100**. Write down all the <u>even</u> numbers.

....

....

5 **a** Write all the <u>odd</u> numbers in the correct place on the number grid.

70	71	72	73	74	75
76					

b What did you notice?

....

Given a number, identify one more

1 For each of the following, write down the number which is <u>one</u> more.

⑧

a One more is? [ ]

⑩

b One more is? [ ]

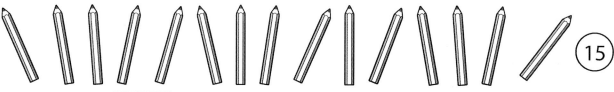

⑮

c One more is? [ ]

2 Complete the following. One is done for you.

1 more than **4** is [5]

a **1** more than **7** is [ ]

b **1** more than **9** is [ ]

c **1** more than **36** is [ ]

d **1** more than **79** is [ ]

e **1** more than **100** is [ ]

f **17** is **1** more than [ ]

g **100** is **1** more than [ ]

Name .... Class .... Date ....

Given a number, identify one less

1 For each of the following, write down the number which is <u>one</u> less.

(7)

a One less is? [   ]

(12)

b One less is? [   ]

(24)

c One less is? [   ]

---

2 Complete the following. One is done for you.

1 less than **3** is [2]

a **1** less than **6** is [   ]

b **1** less than **21** is [   ]

c **1** less than **14** is [   ]

d **1** less than **38** is [   ]

e **1** less than **100** is [   ]

f **61** is **1** less than [   ]

g **80** is **1** less than [   ]

Name .... Class .... Date ....

Identify and represent numbers using objects and pictorial representations including the number line

1 Write the missing numbers on the t-shirts.

2 Write the missing numbers on the bottles.

3 Complete the number lines. Look carefully.

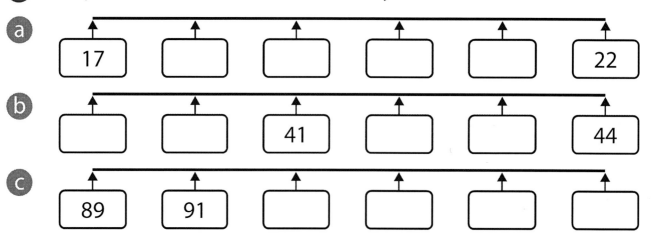

a 17 ☐ ☐ ☐ ☐ 22

b ☐ ☐ 41 ☐ ☐ 44

c 89 91 ☐ ☐ ☐ ☐

4 Now try and complete the missing numbers in the number grid. Part of the grid has been torn off.

0	1	2	3	4	5	6	7	8	9
10	11								19
20				24					
							37		
							47		

Name .... Class .... Date ....

Count and compare numbers up to 100, using the language of: equal, more than, less than (fewer), most, least

1 Write <u>equal</u>, <u>more</u> or <u>fewer</u> in each of the sentences below.

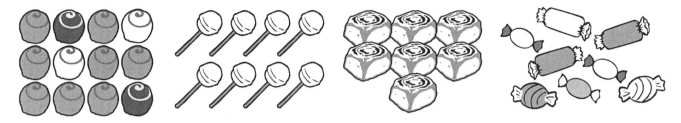

a There are [   ] lollipops than chocolates.

b The number of sweets is [   ] than the number of buns.

c There are [   ] buns than chocolates.

d The number of lollipops is [   ] to the number of sweets.

2 Write <u>more than</u>, <u>less than</u> or <u>equal to</u>.

a • • • • • • is [   ] 14
• • • • • •
• • • • • •
• • • • • •

b ✕✕✕✕✕✕✕✕✕✕
✕✕✕✕✕✕✕✕✕✕
✕✕✕✕✕✕✕✕✕✕ is [   ] 44
✕✕✕✕✕✕✕✕✕✕
✕✕✕✕

c 47 is [   ] 52

d 84 is [   ] 75

e IIIIIIIIII IIIIIIIIII IIIIIIIIII IIIIII
IIIIIIIIII IIIIIIIIII IIIIIIIIII
IIIIIIIIII IIIIIIIIII IIIIIIIIII is [   ] 96

Name .... Class .... Date ....

Recognise repeating patterns with objects and shapes

1 Which toy comes next? Draw or write your answers.

a

  doll   drum   teddy   doll   drum   teddy

b

  car   rocket   rocket   car   rocket   rocket

2 Draw the next **three** shapes in the pattern.

a ○ □ □ ○ □ □ ○

b △ ○ ○ □ △ ○ ○

3 What are the missing toys or shapes? Draw or write your answers.

a

  bike   bus   bike   bus   plane

b □ △   □ △ ○

  square   triangle   square   triangle   circle

c □ □ △ ○ □ △

       square   triangle   circle   square   triangle

Name .... Class .... Date ....

Read and write numbers from 1 to 20 in numerals and words

1 Match up the numerals to the words. One is done for you.

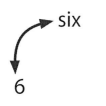

six   ten   twelve

6   15   12   4   20   10

fifteen   four   twenty

2 Write the numerals for the following. One is done for you.

thirteen [13]

a three [   ]   **d** seventeen [   ]

b two [   ]   **e** eleven [   ]

c eight [   ]   **f** nineteen [   ]

3 Now, try and write these numbers in words.

a 7 ....

b 11 ....

c 14 ....

d 16 ....

e 20 ....

Number – number and place value   **YEAR 1**

Solve mixed problems involving the number system

1 Count the toys. How many are there altogether? ⬜

2 Zac was counting his cars. He counted **twelve** in his garage, then another **7** on the floor.

What number did Zac count to ? ⬜

3 **a** Tasmin was counting in **fives**. She said ⑤, ⑩, ⑮.
Write down the next **6** numbers she said.

⬜ ⬜ ⬜ ⬜ ⬜ ⬜

b Now go back and put a circle around all the <u>even</u> numbers which Tasmin said.

4 Put a number on each train carriage so that they go backwards in **twos**.

Continued overleaf

5 Count the sweets and write how many there are in numerals and words.

[ ]

....

---

6 Complete each sentence below by writing <u>odd</u> or <u>even</u>.

a When I start at **0** and count in **twos**, all the numbers are ....

b When I start at **11** and count in **twos**, all the numbers are ....

c When I start at **0** and count in **tens**, all the numbers are ....

d When I start at **5** and count in **twos**, all the numbers are ....

---

7 Count forwards in **twos**. The first number you say is **20**.

What is the **6th** number you say? [ ]

---

8 Count backwards in **fives**. The first number you say is **40**.

What is the **third** number you say? [ ]

HeadStart
Primary

Number

Addition and subtraction

YEAR 1

Name .... Class .... Date......................

Read and write mathematical statements involving addition (+), subtraction (−) and equals (=) signs

1 Draw lines from each sign to its meaning.

| − | + | = |

add   equals   subtract

2 Fill in each box by writing **+**, **−** or **=**. One is done for you.

a and $4 \boxed{} 2 \boxed{=} 6$

b and $9 \boxed{} 7 \boxed{} 16$

c take away $9 \boxed{} 3 \boxed{} 6$

3 Now try these.

a $6 \boxed{} 3 = 9$ **d** $6 \boxed{} 2 + 4$

b $8 \boxed{} 5 = 3$ **e** $11 = 7 \boxed{} 4$

c $7 - 4 \boxed{} 3$ **f** $15 \boxed{} 9 = 6$

Name .... Class .... Date ....

Interpret mathematical statements involving addition (+), subtraction (–) and equals (=) signs

1 Solve the following. Use the pictures to help you.

a) $8 + 5 = \boxed{}$

b) $\boxed{} = 5 + 8$

c) $13 - 5 = \boxed{}$

d) $13 - \boxed{} = 5$

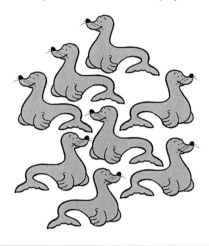

2 Now use the dots to help you solve these.

a) $4 + 6 = \boxed{}$

b) $10 - 4 = \boxed{}$

c) $9 - 2 = \boxed{}$

d) $2 + \boxed{} = 9$

3 Now try these. You could draw your own dots to help you.

a) $3 + 2 = \boxed{}$

b) $\boxed{} = 5 + 3$

c) $6 - 2 = \boxed{}$

d) $7 - 4 = \boxed{}$

Represent and use number bonds and related subtraction facts within 10

1 Draw the correct number of dots in each empty box.

a    **c**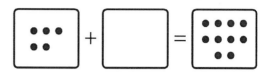

b   **d**

2 Now write the missing numbers in the boxes, so that each pair makes the number in the circle.

a 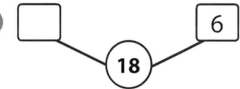   **c**

b 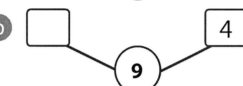   **d**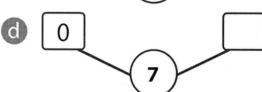

3 Complete the number sentences on each petal to make **10**.
Use different numbers on each petal.

a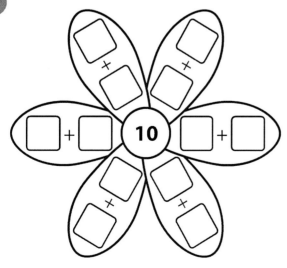

b Use the information to help solve:

$10 - 6 = \boxed{}$

$10 - 3 = \boxed{}$

$10 - \boxed{} = 8$

$10 - \boxed{} = 10$

Number – addition and subtraction   **YEAR 1**

Represent and use number bonds and related subtraction facts within 20

1 Write down as many different ways as you can to make **20** by adding **2** numbers together. Use different numbers each time.

☐ + ☐ = 20   ☐ + ☐ = 20   ☐ + ☐ = 20

☐ + ☐ = 20   ☐ + ☐ = 20   ☐ + ☐ = 20

☐ + ☐ = 20   ☐ + ☐ = 20   ☐ + ☐ = 20

☐ + ☐ = 20   ☐ + ☐ = 20

2 Write the missing numbers so that each pair makes the number in the centre.

a

☐ + 0

☐ + 4   9 + ☐

13

12 + ☐   2 + ☐

3 + ☐

b Use the information to help solve:

$13 - 10 = $ ☐

☐ $- 0 = 13$

$13 - 9 = $ ☐

$13 - $ ☐ $= 1$

3 Put the missing numbers in the boxes.

a ☐ $+ 16 = 20$ and $20 - $ ☐ $= 16$

b $18 + $ ☐ $= 18$ and $18 - 18 = $ ☐

c $19 - 4 = $ ☐   and $4 + $ ☐ $= 19$

Name .... Class .... Date ....

Add one-digit numbers to 20, including zero

1 Solve the following. You can use the dots to help you.

a) 3 + 4 = ☐
● ● ●   : :

c) 2 + 8 = ☐
● ●   ● ● ● ●
       ● ● ● ●

b) 2 + 0 = ☐
● ●

d) 7 + 5 = ☐
● ● ● ●   ● ● ● ●
: : :   ●
● ● ●

2 Now try these. You can draw your own dots to help you.

a) 2 + 6 = ☐

d) 4 + 7 = ☐

b) 4 + 0 = ☐

e) ☐ = 5 + 8

c) ☐ = 3 + 3

f) 6 + 9 = ☐

3 Crack the code. Solve each calculation and then write the letter.

a	b	e	i	m	n	p	r	u	z
19	6	17	14	15	9	13	16	18	4

a) 6 + 3 = ☐

d) 6 + 0 = ☐

b) 9 + 9 = ☐

e) 8 + 9 = ☐

c) 9 + 6 = ☐

f) 8 + 8 = ☐

What word have you made? ☐ ☐ ☐ ☐ ☐ ☐

Name .... Class .... Date ....

Add one-digit and two-digit numbers to 20, including zero

1 Solve each calculation.

a) $2 + 12 = \boxed{}$ e) $\boxed{} = 11 + 7$

b) $14 + 0 = \boxed{}$ f) $15 + 3 = \boxed{}$

c) $10 + 10 = \boxed{}$ g) $6 + 11 = \boxed{}$

d) $\boxed{} = 13 + 4$ h) $0 + 17 = \boxed{}$

2 Put a ring around each calculation which equals the number in the box.

EXAMPLE:

| **16** | (3 + 13) | 7 + 10 | (8 + 8) | (12 + 4) | 10 + 5 |

a) | **19** | 11 + 8 | 4 + 16 | 17 + 2 | 14 + 5 | 13 + 6 |

b) | **17** | 17 + 0 | 7 + 12 | 11 + 6 | 14 + 4 | 14 + 3 |

c) | **18** | 14 + 6 | 12 + 6 | 13 + 5 | 17 + 2 | 19 + 0 |

d) | **14** | 0 + 13 | 1 + 13 | 14 + 0 | 11 + 3 | 12 + 3 |

e) | **15** | 7 + 12 | 11 + 4 | 12 + 4 | 10 + 5 | 13 + 2 |

Subtract two one-digit numbers, including zero

1 Tom ate **4** chocolates.

How many were left? ▢

2 Solve the following. The dots can help you.

a 8 – 2 = ▢   **b** 9 – 5 = ▢

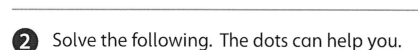

3 Now try these. You can draw your own dots to help you.

a 7 – 4 = ▢   **c** ▢ = 9 – 2

b 6 – 0 = ▢   **d** 8 – 8 = ▢

4 Complete each calculation, and then circle those which have an answer of less than **5**. One is done for you.

7 – 6 = 1

a 3 – 0 = ▢   **c** 8 – 2 = ▢   **f** 7 – 7 = ▢

b 7 – 1 = ▢   **d** 9 – 2 = ▢   **g** 7 – 3 = ▢

                         e 8 – 4 = ▢   **h** 9 – 4 = ▢

Name .... Class .... Date ....

Subtract one-digit and two-digit numbers, including zero

1 Solve the following.

(a) $18 - 6 = \boxed{}$

(b) $17 - 5 = \boxed{}$

(c) $\boxed{} = 19 - 12$

(d) $\boxed{} = 17 - 13$

(e) $14 - 6 = \boxed{}$

(f) $15 - 8 = \boxed{}$

(g) $\boxed{} = 12 - 0$

(h) $10 - 10 = \boxed{}$

(i) $13 - 9 = \boxed{}$

(j) $28 - 15 = \boxed{}$

2 Answer the calculations on the leaves. Then follow the instruction on the flower.

(a)

Shade the leaf if the answer is **even**

$17 - 13 = \boxed{}$

$18 - 5 = \boxed{}$

$19 - 4 = \boxed{}$

$19 - 11 = \boxed{}$

$12 - 8 = \boxed{}$

(b)

Shade the leaf if the answer is **more than 6**

$18 - 12 = \boxed{}$

$17 - 17 = \boxed{}$

$19 - 1 = \boxed{}$

$14 - 5 = \boxed{}$

$15 - 6 = \boxed{}$

Solve missing number problems

1 Put the missing numbers in the boxes.

a) [ ] + 8 = 9   f) 8 + [ ] = 8

b) 4 + [ ] = 17   g) [ ] – 4 = 13

c) 19 – [ ] = 11   h) 16 – [ ] = 7

d) [ ] – 9 = 0   i) [ ] + 13 = 19

e) 6 + [ ] = 17   j) [ ] – 12 = 3

2 Complete the calculations on the carriages so that each answer equals the number on the engine. One is done for you.

a)

13 | 6 + 7 | 19 – [ ] | [ ] – 2 | 10 + [ ]

b)

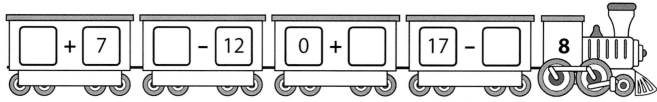

[ ] + 7 | [ ] – 12 | 0 + [ ] | 17 – [ ] | 8

c) Think carefully about this one! Don't write any numbers greater than **20**.

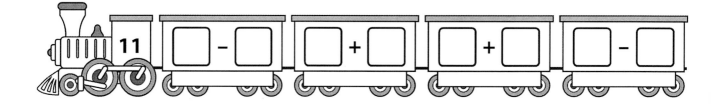

11 | [ ] – [ ] | [ ] + [ ] | [ ] + [ ] | [ ] – [ ]

Name .... Class .... Date ....

Solve one-step problems that involve addition

1 How many animals are there altogether? [ ]

2 Saima and Josh each baked **5** cakes. They put them together on a plate. How many cakes were on the plate? [ ]

3 What is the total cost of the lollipop and the crisps? [ ] p

4 The height of the bigger box is **13** centimetres.
The height of the smaller box is **7** centimetres.

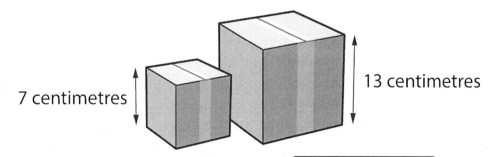

7 centimetres

13 centimetres

What is the total height of both boxes? [centimetres]

Continued overleaf

Name .... Class .... Date ....

5 Deri put **11** spoonfuls of sugar in a bowl.
Then she added **4** more spoonfuls.
How many spoonfuls of sugar did she
put in the bowl altogether?

6 **12** more fish were put in the fish tank. What was
the total number of fish in the tank then?

7 How much money is there in total? [   ] p

8   Jack's cars   Polly's cars

   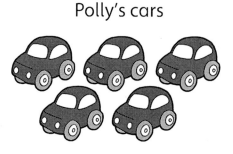

Aftab had **4** cars. How many cars did
Jack, Polly and Aftab have altogether?

Solve one-step problems that involve subtraction

1 **4** of the frogs hopped out of the pond. How many frogs were left in the pond? ☐

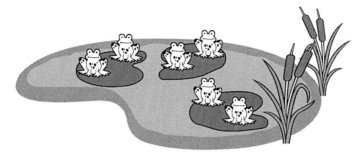

2 **7** of the kites flew away. How many kites were left? ☐

3 There are **8** fewer cows than sheep. How many cows are there? ☐

4 What is the distance between point **A** and point **B**? ☐ centimetres

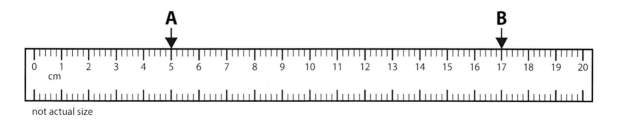

not actual size

Continued overleaf

Name .... Class .... Date......................

5 What is the difference between the number of dragons and the number of dolls? [ ]

6 Each child eats **one** biscuit. How many biscuits are left? [ ]

7 The lollipop cost **9p** less than the carton of drink. [p]
How much does the lollipop cost?

8 Jaz has **17p**. He buys a bun. How much does he have left? [p]

Name .... Class .... Date ....

Solve mixed one-step problems involving addition and subtraction (Choose the correct operation)

1 There were **14** blue cubes and **3** red cubes. How many cubes were there altogether?

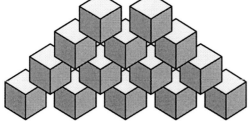

2 Ayaan thought of the number **6**. He added **8** to it.

What was his answer?

3 What is the difference between the length of the bus and the length of the bicycle?

metres

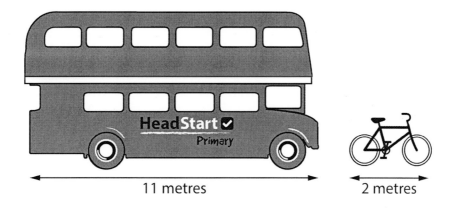

11 metres   2 metres

4 Jai had **20p**.   Jessica had **2p**.

How much more money did Jai have than Jessica?   p

Continued overleaf

Name .... Class .... Date ....

5 Sky had **18** toy cars. Salma had **14** toy cars.

How many more toy cars did Sky have than Salma?

Sky's cars

Salma's cars

6 How much money is there altogether? [ ] p

7 Look at the plan of the school corridor (not actual size).

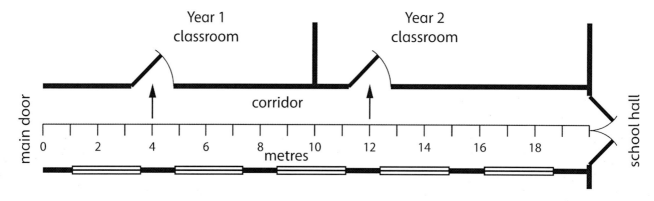

a What is the distance between the Year 1 classroom door and the Year 2 classroom door? [ ] metres

b Samir skipped from the main door to the Year 2 door. Then he skipped another **5 metres**.

How many metres had he skipped altogether? [ ] metres

8 Toby has **2** more blue cubes than red cubes. He has **5** red cubes. How many cubes does Toby have? [ ]

Primary

Number

Multiplication and division

YEAR 1

Solve one-step multiplication problems using pictorial representation. Identify the grouping

1

Anya's toys    Alia's toys

[ ] groups of **2** toys equal [ ] toys in total.

2 For each of the following, fill in the missing numbers.

4 groups of [ ] flowers make [ ] flowers altogether.

3

[ ] lots of **5** sweets equal [ ] sweets altogether.

4

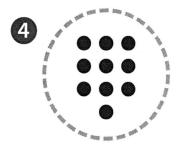   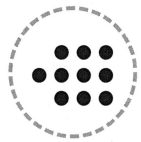

[ ] groups of **10** spots equal [ ] spots.

Continued overleaf

Name .... Class .... Date ....

Think carefully about these. You need to fill in all the missing numbers in the sentences.

[ ] lots of [ ] cakes make [ ] cakes altogether.

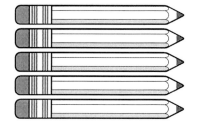 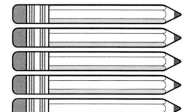 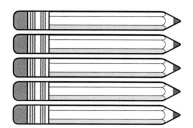

[ ] groups of [ ] pencils equal [ ] pencils.

[ ] lots of [ ] apples make [ ] apples in total.

  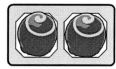

[ ] groups of [ ] chocolates equal [ ] chocolates altogether.

Solve one-step multiplication problems using pictorial representations

1 How many gloves are there altogether? ☐

Count in **twos** to find the answer.

2 Count in **tens** to work out the total number of spots. ☐

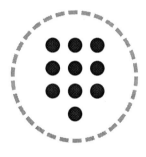  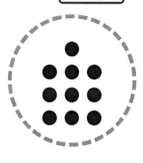

3 How many sweets are there altogether? ☐

Count in **fives** to get the answer.

4 How many stars are there on the t-shirts altogether? ☐

Count in **twos**.

Continued overleaf

5 How many animals are there altogether? ⬜

6 Each child has **2** sweets. How many sweets are there altogether? ⬜

7 Each packet contains **10** pencils.
How many pencils are there altogether? ⬜

10 pencils | 10 pencils | 10 pencils | 10 pencils | 10 pencils | 10 pencils | 10 pencils | 10 pencils | 10 pencils

8 Each bunny eats **5** carrots.
What is the total number of carrots eaten? ⬜

Solve one-step multiplication problems using arrays. Identify the grouping

For each of the following, fill in the missing numbers.

1 **3** rows of ☐ lollipops

make ☐ lollipops.

2

☐ rows of **10** spots equal ☐ spots altogether.

3 **4** lots of ☐ frogs make ☐ frogs in total.

4 ☐ lots of **8** dog bones total ☐ dog bones.

Continued overleaf

Name .... Class .... Date ....

For these, fill in all the missing numbers in the sentences.

5 [ ] rows of [ ] dogs make [ ] dogs.

6 [ ] lots of [ ] cows equal [ ] cows.

7

[ ] lots of [ ] circles total [ ] circles.

8 Think really carefully about this one. There are <u>two</u> sentences to complete.

✗ ✗ ✗ ✗ ✗
✗ ✗ ✗ ✗ ✗
✗ ✗ ✗ ✗ ✗

[ ] lots of [ ] crosses make [ ] crosses.

[ ] lots of [ ] crosses make [ ] crosses.

Number – multiplication and division   **YEAR 1**

Solve one-step multiplication problems using arrays

1 There are ☐ rows of dots.

There are ☐ dots in each row.

How many dots are there altogether? ☐

2 There are ☐ chocolates in each row.

There are ☐ rows.

How many chocolates are there altogether? ☐

3

There are ☐ rows of ladybirds. There are ☐ ladybirds in each row.

How many ladybirds are there in total? ☐

4 Look carefully at how many ice-cream vans are in each row, and how many rows there are.

How many ice-cream vans are there altogether? ☐

Continued overleaf

5

How many squares are there altogether?

6

How many footballs are there?

7

How many birds can you see?

8

How many monkeys are hanging from the branches in total?

Name .... Class .... Date ....

Draw and use arrays to solve one-step multiplication problems

1 Franco had **4** jars with **5** marbles in each jar. He made the array below to help him work out how many marbles he had altogether.

What was Franco's answer? ☐

2 There were **7 pairs** of swings. Draw an array to help you work out how many swings there were altogether.

What was the total number of swings? ☐

3 Ben had **3** packets of pencils. There were **10** pencils in each packet. Draw an array to help you work out the total number of pencils.

10 pencils

How many pencils were there altogether? ☐

Name .... Class .... Date ....

Solve one-step multiplication word problems

You can use diagrams or arrays to help you solve these problems.

1 There were **6** drums. Each drum had **2** drumsticks.

What was the total number of drumsticks? ☐

2 There were **10** toy boxes. Each toy box had **5** toys.

How many toys were there altogether? ☐

3 Anna, Freddie and Samir each had **5** sweets.

How many sweets did they have altogether? ☐

4 A café had **10** tables. There were **4** chairs at each table.

How many chairs were in the café in total? ☐

Continued overleaf

5 A purse contained **seven 2 pence** coins.

How much money was there altogether? [p]

6 There were **8** washing lines. **5** sweaters were drying on each line.

How many sweaters were drying altogether? [ ]

7 There were **12** bicycles in the shop.

What was the total number of wheels? [ ]

8 Mrs Egg, the baker, made **10** cakes. Look at the ingredients for **1** cake.
Use them to complete the ingredients for **10** cakes.

To make 1 cake	To make 10 cakes
3 eggs	[ ] eggs
1 glass of milk	[ ] glasses of milk
4 spoonfuls of sugar	[ ] spoonfuls of sugar
2 cups of flour	[ ] cups of flour

SELF RAISING
FLOUR

Name .... Class .... Date ....

Double numbers and quantities

1 For each of the following, draw double the number of carrots in the box.

a

b

c

2 Put the numbers through the doubling machine. One is done for you.

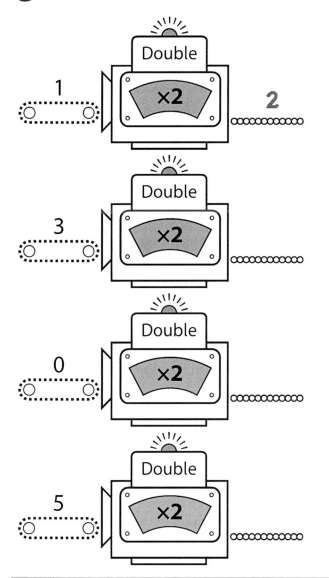

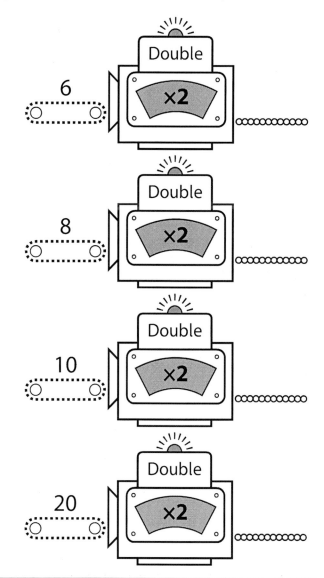

Solve problems using doubling

1  Stefan's apples

Leroy had **double** the number of apples as
Stefan had. How many apples did Leroy have?

2 There are **twice** as many buses as rockets.

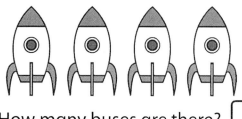 

How many buses are there?

3 Zoe had **9p**. Zain had **double** that amount.

How much money did Zain have? [ ] p

4 There are **twice** as many boys as girls
on the park. There are **12** girls.

How many boys are there?

5 The height of the doll's chair is
double the height of the doll's table.

What is the height of the doll's chair?

[centimetres]

20 centimetres

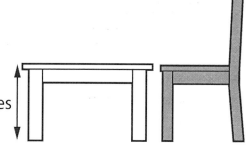

Name .... Class .... Date ....

Solve one-step division problems using pictorial representations. Identify the grouping

Fill in the missing numbers in the sentences.

1

10 teddies shared between ☐ equal ☐ teddies in each group.

2

12 sweets shared between ☐ bags give ☐ sweets in each bag.

3 **20** straws put into ☐ milk bottles give ☐ straws in each bottle.

4

9 toy boats shared between ☐ groups equal ☐ toy boats in each group.

Continued overleaf

For the following, use the objects to complete all the numbers in the sentences.

5

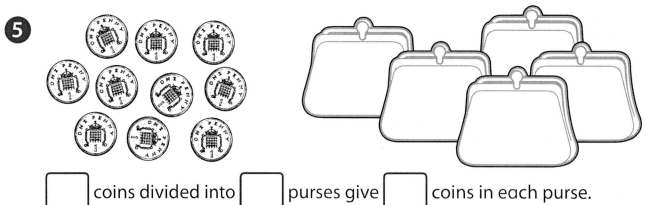

☐ coins divided into ☐ purses give ☐ coins in each purse.

6

☐ candles shared between ☐ cakes give ☐ candles on each cake.

7

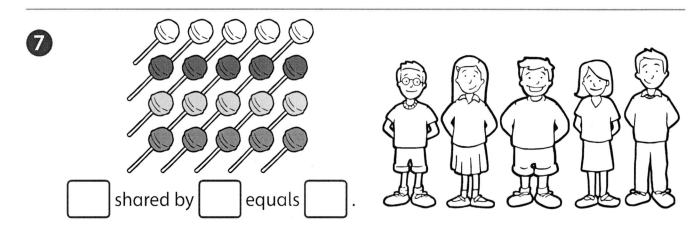

☐ shared by ☐ equals ☐ .

8

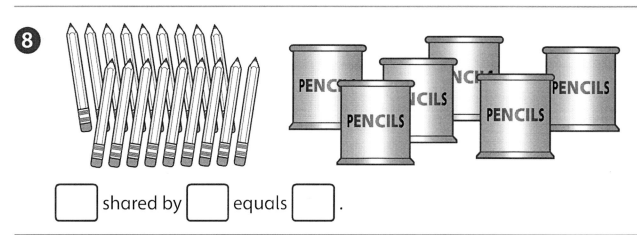

☐ shared by ☐ equals ☐ .

Name .... Class .... Date ....

Solve one-step problems involving division using pictorial representations

1 Caitlin put the same number of cakes on each plate.

 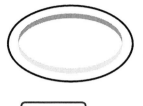 

How many cakes were on each plate? ☐

2 Tom had **30** cubes. He made **10** towers. Each tower had the same number of cubes.

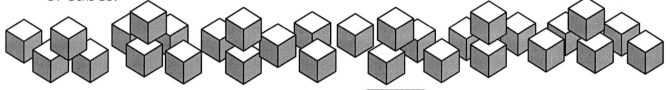

How many cubes were in each tower? ☐

3 **Five** friends shared **15** balls evenly between them.

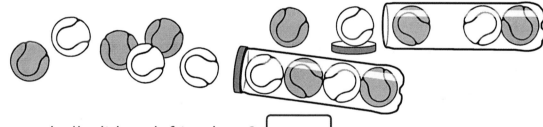

How many balls did each friend get? ☐

4 Put the same number of dots in each circle so that there are **16** altogether.

Continued overleaf

5 The flower shop had **9** flowers. They shared them evenly between **3** vases.

How many flowers were in each vase? [ ]

6 Mrs Shah had **12** sweets. She shared them evenly between **4** children.

How many sweets did each child get? [ ]

7 A piece of rope measured **16 metres**. It was cut evenly into **4** pieces.

How many metres was each piece? [metres]

8 **18** chocolates were shared evenly between **3** boxes.

How many chocolates were in each box? [ ]

Name .... Class .... Date ....

Solve one-step division problems using arrays

1 **15** chocolates shared between

5 rows give [ ] in each row.

2 **16** cubes shared between **2** is [ ] .

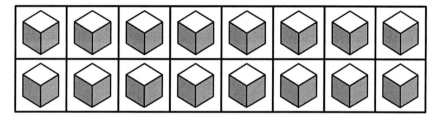

3 **30** shared by **10** is [ ] .

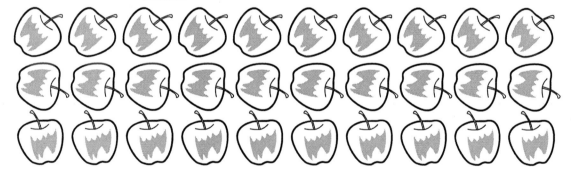

4 **35** shared by **5** equals [ ] .

Continued overleaf

5 **12** shared by **4** is ⬚ .

6 **16p** shared by **4** is ⬚ p .

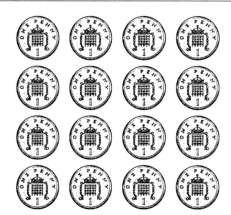

7 **24** shared by **3** is ⬚ .

8 **30** shared by **6** equals ⬚ .

Number – multiplication and division   **YEAR 1**

Solve one-step division problems using arrays. Understand that two problems can be solved with one array.

For each of the following, solve **2** problems using the pictures.

1   **15** shared by **3** equals [ ] ,

and **15** shared by **5** equals [ ] .

2   **10** shared by **2** is [ ] ,

and **10** shared by **5** is [ ] .

3   **30** shared by **2** equals [ ] and **30** shared by **15** equals [ ] .

✗ ✗ ✗ ✗ ✗ ✗ ✗ ✗ ✗ ✗ ✗ ✗ ✗ ✗ ✗
✗ ✗ ✗ ✗ ✗ ✗ ✗ ✗ ✗ ✗ ✗ ✗ ✗ ✗ ✗

4   **20** shared by **5** equals [ ] ,

and **20** shared by **4** equals [ ] .

Continued overleaf

For these, write the missing numbers to make **2** different number sentences.

5 **12** shared by ☐ is ☐. **12** shared by ☐ is ☐.

6 **24** shared by ☐ is ☐. **24** shared by ☐ is ☐.

7 **14** shared by ☐ equals ☐. **14** shared by ☐ equals ☐.

8 ☐ shared by ☐ is ☐.

   ☐ shared by ☐ is ☐.

Name .... Class .... Date ....

Draw and use arrays to solve one-step division problems

1 Annie had **15** sweets. She wanted to share them between **5** children. She made this array to help her work out how many sweets to give each child. She counted the dots in **fives** until she got to **15**.

1	2	3	4	5
●	●	●	●	●
6	7	8	9	10
●	●	●	●	●
11	12	13	14	15
●	●	●	●	●

Annie's friend made the array like this.

1 ●   6 ●   11 ●

2 ●   7 ●   12 ●

3 ●   8 ●   13 ●

4 ●   9 ●   14 ●

5 ●   10 ●   15 ●

How many sweets did each child get? ⬜

2 There were **12** fish in **3** fish bowls.

Draw an array to help you work out how many fish were in each bowl.

How many fish were in each bowl? ⬜

3 What is **21** shared by **7**? ⬜

Draw an array to help you work out **21** shared by **7**.

4 What is **16** shared by **4**? ⬜

Draw an array to help you work out **16** shared by **4**.

Name .... Class .... Date.......................

Solve one-step division word problems

You can use diagrams or arrays to help you solve these problems.

1 Dylan and Ali shared the cars evenly.

How many did they each get? ☐

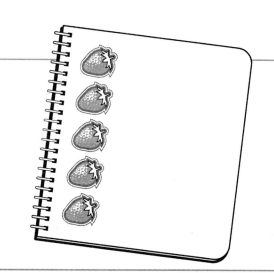

2 Jon had **20** stickers.
He put them into **5** even groups.

How many stickers
were in each group? ☐

3 There are **70** bricks.

How many towers of
10 bricks can be made? ☐

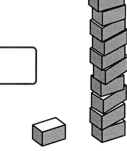
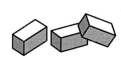

4 Year 1 get **28** new books.
They divide them evenly between **4** shelves.

How many books do
they put on each shelf? ☐

Continued overleaf

Name .... Class .... Date ....

5 There are **30** children in Year 1. They are divided into **5** even groups.

How many children are in each group? ☐

6 **3** sweets cost **18p** altogether. Each sweet costs the same amount.

How much does one sweet cost? [p]

7 The school garden was divided evenly into **4** sections.

How long was each section? [metres]

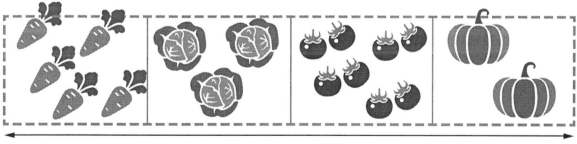

24 metres

8 Ryan saved the same amount every day for **3** days. He saved **24p** altogether.

How much did he save each day? [p]

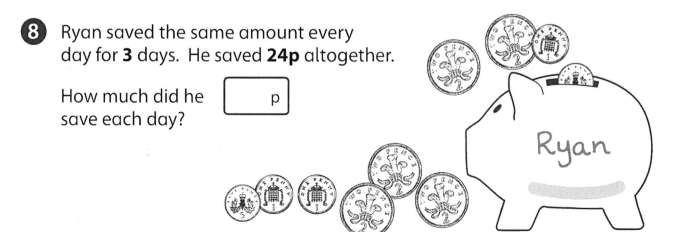

Name .... Class .... Date ....

Find simple fractions of objects, numbers and quantities

1 For each of the following, put a ring around **half** of the animals.

a

b

2 Now put a ring around a **quarter** of the shapes.

a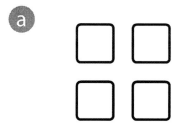

b

3 Complete the following.

a A **half** of 4 is [ ].

b A **half** of 10 is [ ].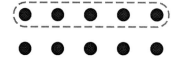

c A **quarter** of 12 is [ ].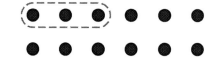

d A **half** of 16 is [ ].

You could draw your own dots to help you.

e A **quarter** of 20 is [ ].

Solve problems involving finding fractions

1 **Half** of the frogs jumped out of the pond.

How many frogs jumped out? ☐

2 **Half** of the shirts fell off the line.

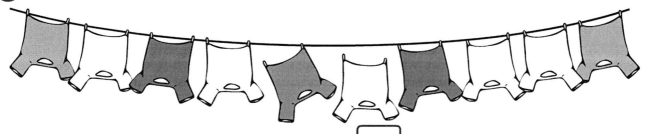

How many shirts were still on the line? ☐

3 Isaac had **20p**. Jayden had **half** as much as Isaac.

How much did Jayden have? ☐ p

4 The rabbit ate **one quarter** of the carrots.

How many carrots did the rabbit eat? ☐

 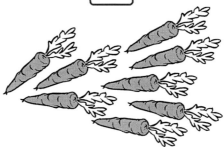

Continued overleaf

Name .... Class .... Date ....

5 **Half** of the cubes are blue. A **quarter** of the cubes are red.

a How many cubes are blue? ☐ b How many cubes are red? ☐

6 A **quarter** of the pens do not work. How many pens do not work? ☐

7

24 centimetres   (not actual size)

a The pencil is **half** as long as the string.

How long is the pencil? ☐ centimetres

b The toy spoon is a **quarter** as long as the string.

How long is the toy spoon? ☐ centimetres

c The toy knife is one **third** as long as the string.

How long is the toy knife? ☐ centimetres

Solve one-step problems involving multiplication and division.
(Choose the correct operation)

1 May had **2** dolls. Huma had **5** times as many dolls as May.

How many dolls did Huma have? ☐

2 There are **3** cars in the garage. There are **10** times as many cars on the car park as in the garage.

How many cars are on the car park? ☐

3 Mum bought **6** chocolate bars. She shared them evenly between her **2** children. How many chocolate bars did each child get? ☐

CHOCOLATE CHOCOLATE CHOCOLATE CHOCOLATE CHOCOLATE CHOCOLATE

4 Some potatoes weighed **4 kilograms**.

a Some apples weighed **half** as much as the potatoes.

How much did the apples weigh? ☐ kilograms

b Some tomatoes weighed **4 times** as much as the potatoes.

How much did the tomatoes weigh? ☐ kilograms

Continued overleaf

5 Hannah is **7 years old**. Her brother, Harry, is **double** Hannah's age.

How old is Harry? | years old |

6 Mrs Black, the headteacher, buys **20** new paint boxes.
She divides them evenly between **4** classes.

How many paint boxes
does each class get? [ ]

7 Each tower has **6** cubes. There are **4** towers.

What is the total number of cubes? [ ]

8

3p

12p

a Lollipops cost **half** as much as buns. How much do lollipops cost? | p |

b How much would **10** lollipops cost? | p |

c Crisps cost **5 times** as much as space ships.

How much do crisps cost? | p |

d How many space ships could you buy with **15p**? [ ]

HeadStart ✓
Primary

Number

Fractions

Name .... Class .... Date ....

Recognise, find and name a half as one of two equal parts of an object

1 **a** Choose the correct word to complete the sentence.

half   quarter   whole

[   ] of the t-shirt is shaded

b What fraction of the t-shirt is not shaded? [   ]

---

2 **Half** of the football is black. The rest is white.

a What fraction of the football is white? [   ]

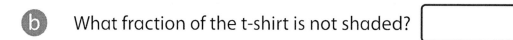

b Circle the correct words in the sentence below.

The white part of the football is **more than/less than/equal to** the black part of the football.

---

3 Shade **half** of the bottle.

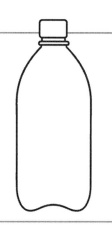 

---

3 Put a tick (✓) by the objects which have **one half** shaded.

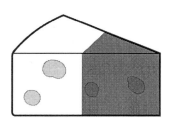  

[ ]   [ ]   [ ]

Recognise, find and name a half as one of two equal parts of a shape

1 Look at the rectangle.

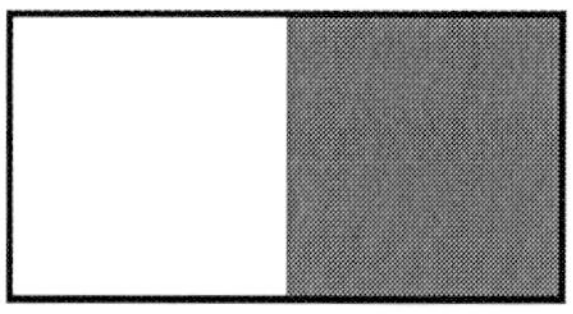

a What fraction of the shape is shaded?

b What fraction of the shape is not shaded?

2 Shade **one half** of each of the shapes.

3 Put a tick (✓)by the shapes, which have **one half** shaded.

a

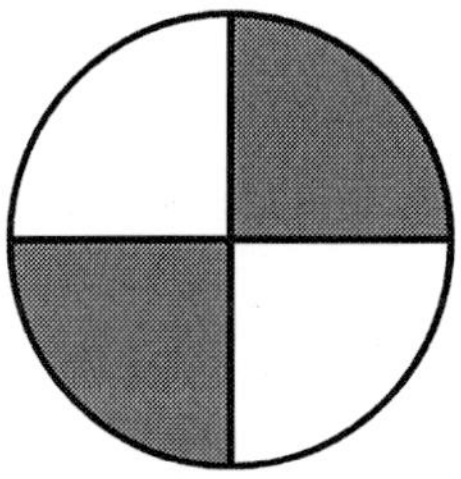

c

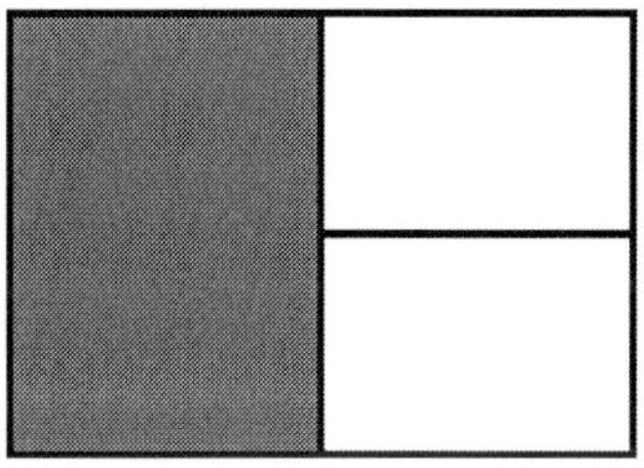

b

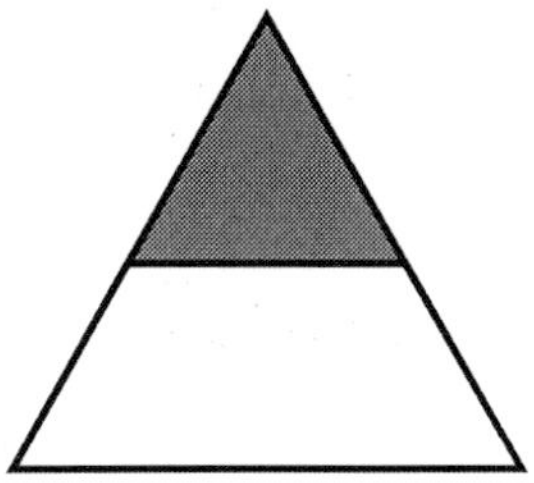

d

4 Complete this sentence.

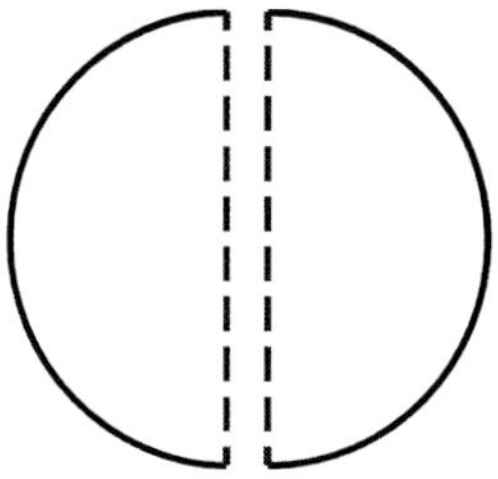

Two halves of a circle make [   ] circle.

Recognise, find and name a half as one of two equal parts of a quantity

1

What fraction of the toys are dolls? ☐

2

What fraction of the toys are trains? ☐

3 Colour **half** of the buttons.

4 Complete the following.

a **Half** of **2** is ☐

b **Half** of **8** is ☐

c **Half** of **12** is ☐

d **Half** of **16** is ☐

e **Half** of **22** is ☐

Drawing your own dots may help you.

Name .... Class .... Date ....

Recognise, find and name a half as one of two equal parts of a length

1

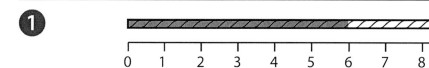

0 1 2 3 4 5 6 7 8 9 10 11 12
centimetres

a What fraction of the rope is shaded? _____

b What fraction of the rope is not shaded? _____

2 Put an arrow (↓) half way along the measure.

0 cm 1   2   3   4   5   6   7   8   9   10

3

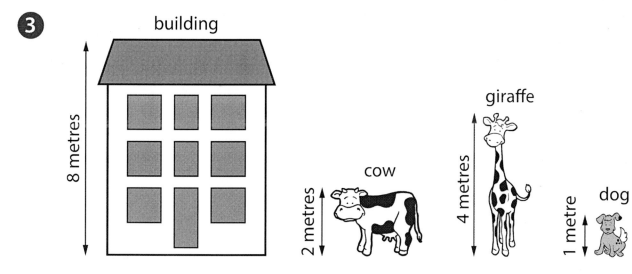

a Which animal is **half** as tall as the building? _____

b Complete the sentence.

The _____ is **half** the height of the giraffe and the _____

is **half** the height of the cow.

Recognise, find and name a quarter as one of four equal parts of an object

1

PIZZA TOPPINGS

grated cheese

mushrooms

sliced pepper

tomatoes

Choose a word to complete each sentence.

quarter   half   equals   whole

a The fraction of the pizza topped with tomatoes [   ] the fraction topped with grated cheese.

b A [   ] of the pizza is topped with mushrooms.

2

a Shade a **quarter** of the jar.

Put dots [•] on a **quarter** of the jar.

Put lines [/] on a **quarter** of the jar.

Put crosses [×] on the rest of the jar.

b What fraction of the jar has crosses [×] on it? [   ]

c How many different parts are there on the jar? [   ]

Name .... Class .... Date ....

Recognise, find and name a quarter as one of four equal parts of a shape

1 What fraction of the shape is shaded?

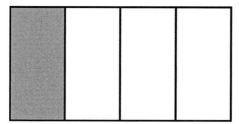 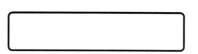

2 Draw more lines on each shape to make **quarters**. (You may need one line or more than one line.)

a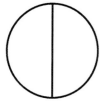

c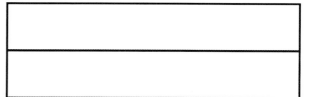

b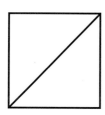

d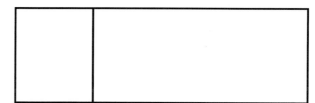

3 Put a tick (✓) next to the shapes which have **one quarter** shaded.

a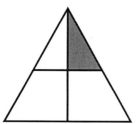

c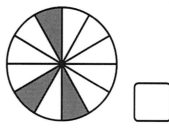

b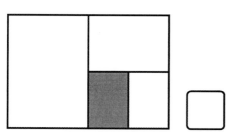

d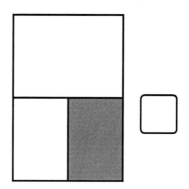

Recognise, find and name a quarter as one of four equal parts of a quantity

1 What fraction of the animals are rabbits?

2 What fraction of the shapes are squares?

3 Put a ring around **one quarter** of the children.

4 Leo breaks the chocolate into **quarters**.

a How many pieces does he have after he has broken the chocolate into **quarters**?

b He gives **one quarter** to Ethan, **two quarters** to Mai and keeps **the rest** for himself.

Ethan says, "My piece is smaller than Leo's piece." Is Ethan correct? Explain your answer.

....

Name .... Class .... Date ....

Recognise, find and name a quarter as one of four equal parts of a length

1 Look at the distances below.

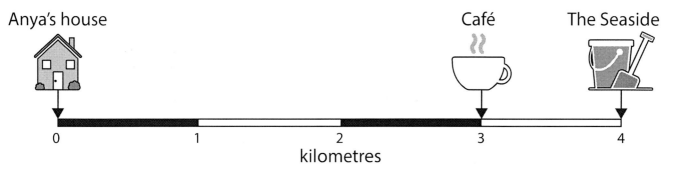

Anya's house   Café   The Seaside

0   1   2   3   4
kilometres

a Anya starts at her house and cycles **a quarter** of the way to the seaside. Put an arrow (↓) to show how far she has cycled.

b Then she cycled to the café. What fraction of the whole journey (Anya's house to the seaside) had she still to cycle?

2 Draw lines across the measure to divide it into **quarters**. The first is done for you.

0 cm 1   2   3   4   5   6   7   8

3 Eva walks **3 metres** along the path to school.

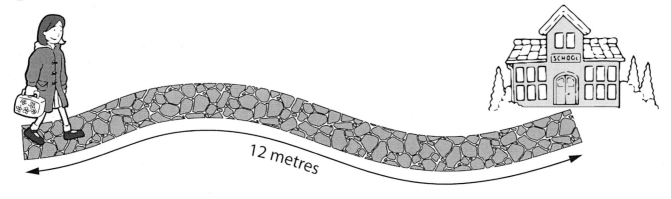

12 metres

What fraction of the path has Eva walked?

Name .... Class .... Date ....

Connect halves and quarters to the equal sharing and grouping of sets of objects and to measures. Recognise and combine halves and quarters as parts of a whole.

1 **Half** of the bicycles ride away. How many are left? ☐

2 Mr Powell has **8** candles.

He puts **half** on Niko's cake.

Draw the candles on the cake.

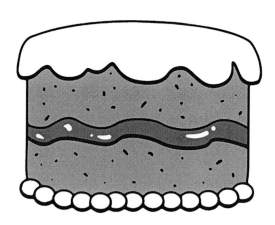

3 What is **one quarter** of **8 metres**? ☐ metres

4 Layla, Lucas and Leroy each get **one quarter** of **£12**

a What fraction is left? ☐

b How much does Layla get? £ ☐

Continued overleaf

Number – fractions   **YEAR 1**

5 Some children swim **16** lengths in total.
They each swim **one quarter**.

a How many children swim? ▢

b How many lengths do they each swim? ▢

c Joe and Jess say, "Together, we have swum one **half** of the total distance."

Are they correct? ▢ Explain your answer.

....

6 There are **12** strawberries in a basket. Ayub eats **one quarter** and Marlon
eats **half** of the strawberries.

a How many strawberries does Ayub eat? ▢

b How many strawberries does Marlon eat? ▢

c Marlon says, " I have eaten **two quarters** of the basket of strawberries."

Is he correct? ▢

Try to explain your answer

....

....

d What fraction of the whole basket
of strawberries is left?

▢

Primary

Measurement

Compare, describe and solve problems for lengths and heights

1 Which is shorter? Put a tick (✓) by your answer.

☐   ☐

2 Which is taller?
Put a tick (✓) by your answer.

☐   ☐

3 Write a word to complete each sentence, so that it compares the **length** or **height**.

a The matchstick is .... than the dog bone.

b The flagpole is .... than the postbox.

4 How many cricket bats would you need to **equal** the length of the wood?

☐

Name .... Class .... Date ....

Compare, describe and solve problems for mass/weight

1 Which animal is **heavier**? Put a tick (✓) by your answer.

 ☐   ☐

2 Put a circle around the **heaviest** and a tick (✓) by the lightest.

 ☐   ☐   ☐

3 Complete each sentence to compare the weights.

Choose from: **lighter than**   **heavier than**   **equal to**

a The toy train is .... the girl.

b The man is .... the girl.

Measurement   **YEAR 1**

Name .... Class .... Date ....

Compare, describe and solve problems for capacity/volume

1 Which jug has more water? Put a tick (✓) by you answer.

2 Draw a line from each glass to the correct word or phrase.

half full   full   quarter full   empty

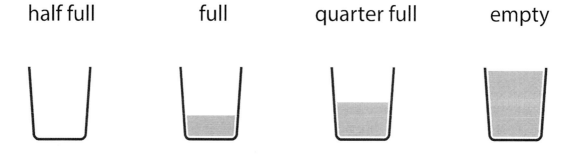

3 Write a word to complete each sentence so that it compares the capacity.
Choose from: **more**   **less**

a The glass has .... water than the vase.

b The bath has .... water than the glass.

                bath   glass   vase

4 The jug was full with water. Then Yannick poured the water into the container.
Which has the greater capacity, the jug or the container?

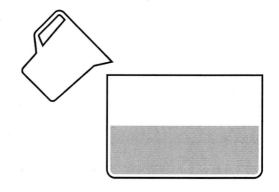

Compare, describe and solve problems for time

1 Which is quicker? Put a tick (✓) by your answer.

pouring a glass of water [ ]   walking around the park [ ]

2 Which happens later in the day? Put a tick (✓) by your answer.

eating breakfast [ ]   walking home from school [ ]

3 Write a word in each sentence to compare the time.

Choose from: **before   quicker   after   slower**

a It is .... to climb **3** stairs than to climb a mountain.

b **8 o'clock** in the morning is .... **6 o'clock** in the evening.

8:00   6:00

4 It took Kira **50 seconds** to swim a length.
It took Kai **44 seconds** to swim a length.

Who was slower? [   ]

Measure and begin to record lengths and heights

1 How many units long is the line? [ ] units

2 How many units high is the gate?

[ ] units

3 Each section of the rope measures one **centimetre** (cm).

How long is the rope? [ ] centimetres

1 cm

not actual size

4 Use your ruler to measure the length of each line.

a

[ ] centimetres

c

[ ] centimetres

b

[ ] centimetres

Name .... Class .... Date ....

Measure and begin to record mass/weight

1 Put a tick (✓) by the scales you could use to measure weight.

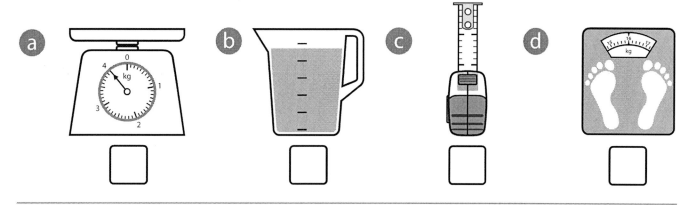

a ☐　　b ☐　　c ☐　　d ☐

2 What is the weight shown on each scale?

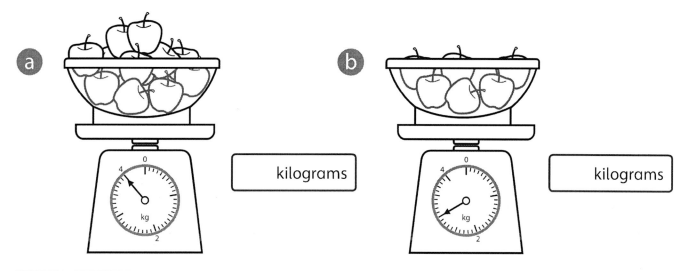

a | kilograms

b | kilograms

3 Now try these.

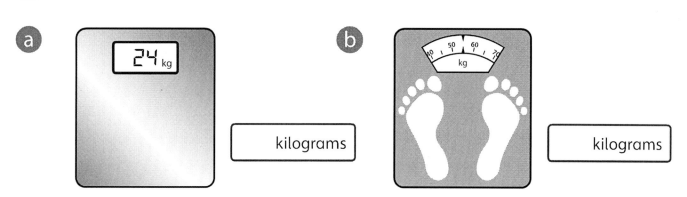

a | kilograms

b | kilograms

Name .... Class .... Date ....

Measure and begin to record capacity/volume

1 Azra was making a jug of fruit punch. Which of the following could she have used to measure the liquid she needed? Tick (✓) the boxes.

a

☐

c

☐

e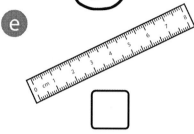

☐

b

☐

d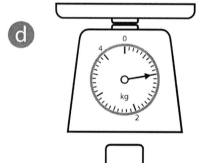

☐

f

☐

2 Write down the amount shown by each arrow. Circle whether your answer is **millilitres** (ml) or **litres** (l). One is done for you.

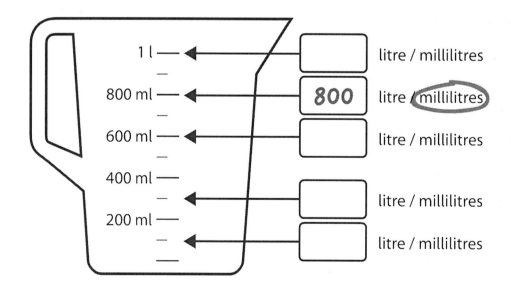

1 l ← [ ] litre / millilitres

800 ml ← [800] litre / (millilitres)

600 ml ← [ ] litre / millilitres

400 ml

← [ ] litre / millilitres

200 ml

← [ ] litre / millilitres

Measure and begin to record time

1 Bethany walks across the classroom. Which of the following amounts of time could it take her? Circle your answer

12 minutes / 12 seconds / 12 hours

2 During the night, Rosa goes to bed for **10 seconds / 10 minutes / 10 hours**. Circle the correct amount of time.

3 On Tuesdays, Mrs Zafar's class learn Maths for **50 seconds / 50 minutes / 50 hours**. Circle the correct amount of time.

4 For the following, you will need to use the words **seconds**, **minutes** or **hours** in your answer.

a How long do you think it would take you to wash your hands and face, and clean your teeth?

b How long do you think it would take you to say, "Good morning, Mrs Topping"?

c Alice starts reading her book at **10 o'clock** in the morning. She finishes it at **11 o'clock** in the morning. How long does Alice read for?

d A show starts at **4 o'clock** in the evening. It ends at **7 o'clock** in the evening. How long does the show last for?

4:00   7:00

Recognise and know the value of different denominations of coins and notes

1 Draw lines to match the money to the value.

£1   £5

10p   5p

1p

£10   50p

2 Put a tick (✓) by the set of coins which could be used to pay for the lollipop.

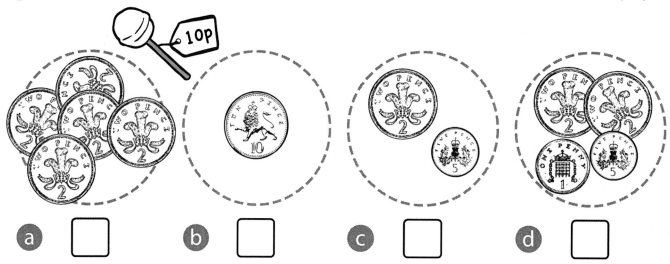

a ☐   b ☐   c ☐   d ☐

3 Draw **2** different sets of coins which you could use to buy the cake.

Sequence events in chronological order using appropriate language

1 Write a word in the spaces so that the story makes sense.

Choose from:

evening   before   morning   first   later   after   next   until

Use each word once.

Abigail's Day

...................., Abigail put on her slippers   . ..., she went

downstairs to eat her breakfast   . ... she got dressed,

Abigail cleaned her teeth   . ... in the ....,

she went to the park   . In the ...., she played with her

friend   , Ayub. .... that, Abigail watched some

television   , ... it was time for bed   .

---

2 Write **today**, **yesterday** or **tomorrow** in the sentences below.

I will be going swimming ....

.... is my birthday.

.... it was raining.

Recognise and use the language related to dates

1 Complete all the days of the week in the correct order, starting with Monday.

Monday
....

....

2 Complete the sentences below. Choose from: **months   weeks   days**

a There are **4** [   ] in one month.

b There are **7** [   ] in one week.

c There are **12** [   ] in one year.

3 Write the correct months in the spaces below.

a The first month of the year is [   ]

b [   ] is the month before May.

c [   ] is the month after August.

d The last month of the year is [   ]

4 Liam wrote **24/10/2015** in his maths book.

a Which number means the day? [   ]

b What was the year? [   ]

c Write the month as a word. [   ]

Name .... Class .... Date ....

Tell the time to the hour and half past hour

1 What time do the clock faces show? Circle your answers.

a

6 o'clock   7 o'clock

12 o'clock   half past 4

b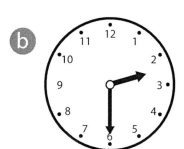

half past 6   6 o'clock

2 o'clock   half past 2

2 Write the correct time beside each clock.

a 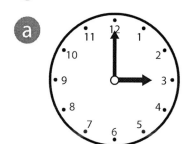 ....

d 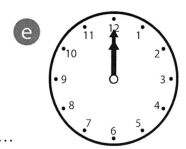 ....

b  ....

e ....

c 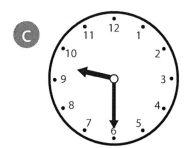 ....

f  ....

Name .... Class .... Date ....

Draw the hands on a clock face to the hour and half past the hour

1 Draw the times on each clock face.

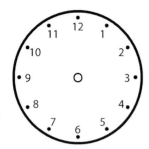 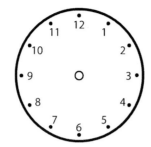 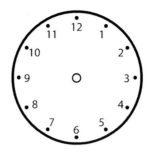 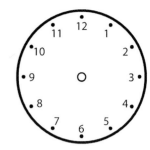

a 7 o'clock **b** 5 o'clock **c** 4 o'clock **d** 10 o'clock

2 Now try these.

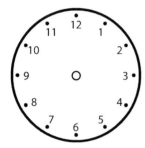 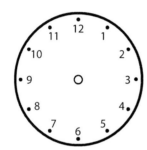 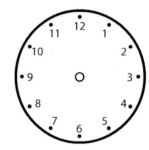 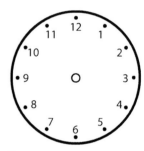

a half past 3 **b** half past 11 **c** half past 5 **d** half past 12

3 Levi went swimming at **6 o'clock**.

a Draw the time on the clock face.

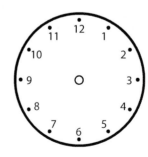

b He got out of the water **half an hou**r later. Draw the time.

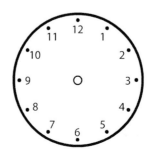

HeadStart ✓
Primary

Geometry

Properties of shapes

YEAR 1

Name .... Class .... Date........................

Recognise and name common 2-D shapes

1 Put a tick (✓) by all the rectangles below.

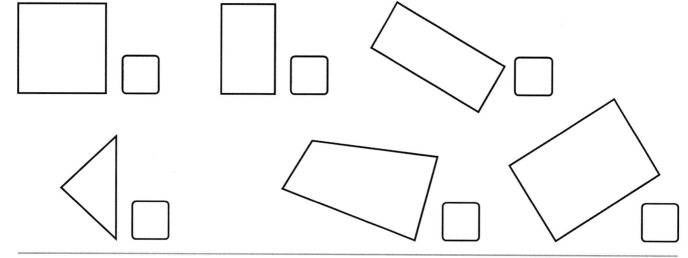

2

What shape is this?

Explain how you know.

....

3 Draw a triangle inside a circle.

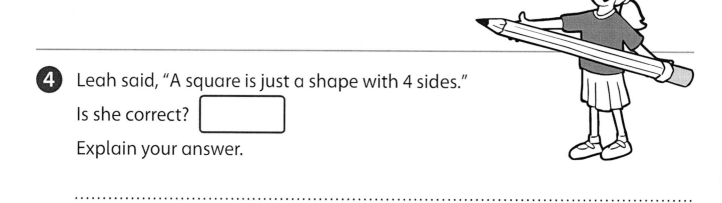

4 Leah said, "A square is just a shape with 4 sides."

Is she correct?

Explain your answer.

....

Name .... Class .... Date ....

Recognise and name common 3-D shapes

1 Draw a line from each **3-D** shape to its name.

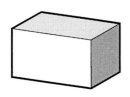 

cube   sphere   cone   cuboid   pyramid

2 Write the name of the shape by each item.

a

c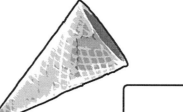

b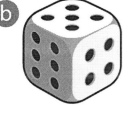

d

3 Max said, ...

A cube is just another name for a cuboid.

Is he correct?

Explain your answer.

....

Geometry

Position and direction

YEAR 1

Describe position, direction and movement using an amount of turn

1 The hour hand is pointing to **12**.

a If it makes **half** a turn, what number will it point to? ☐

b What number will it point to if it makes a **quarter** turn clockwise? ☐

c The hand is pointing to **6**. Anna makes a **quarter** turn anti-clockwise. What number does she turn the hand to? ☐

2 Look at the spinner.

a What will the spinner point to if it makes a **full** turn?

☐

b What will the spinner point to if makes a **quarter** turn anti-clockwise?

☐

trifle

biscuit

cheese

cake

c The spinner turns to point to the cake, after passing the biscuit. Describe its turn.

....

d The spinner is pointing to the cheese. Molly makes a **quarter** turn clockwise, then **half** a turn anti-clockwise. What is the spinner pointing to now?

☐

Name .... Class .... Date ....

Use appropriate language to describe position, direction and movement

1 Look at the position of the vehicles below.

a What is on the top right?

[ ]

bicycle

lorry

b What is on the bottom left?

[ ]

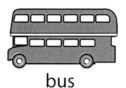
bus

car

c What is on the middle right?

[ ]

ice-cream van

coach

d Describe the position of the bicycle.

The bicycle is on the [ ]

2 Underline the correct words to complete each sentence.

a The drum is **on top of / underneath / beside** the table.

b The skateboard is **on top of / underneath / beside** the table.

c The toy car is **inside / between / outside** the table and the toy box.

d The teddy is **inside / outside/ next to** the toy box.

e The cubes are **on top of / around / inside** the drum.

f The skipping rope is **around / near / above** the skateboard.

Continued overleaf

3 Look at the plan and then put a word in each of the spaces below.

Choose from: **close forwards backwards up far down**

a To get to the house, the car should go ....

b To get to the swings, the car should go ....

c The garden is .... to the swimming pool.

d The pond is .... from the garden.

e To reach the tower, the boy should walk .... the hill.

f The tree is further .... the hill than the tower.

Further mastery

Further mastery – number and place value

1

a Put a tick (✓) by the road which has more cars.

b Count how many cars there are altogether and write your answer in digits and words.

2 Draw the shapes on the **second** necklace so that it is the <u>same</u> pattern as the **first** necklace.

 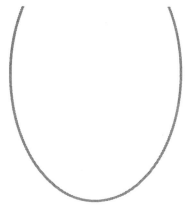

3 **a** Draw this pattern in the boxes below.

The **first** shape is a square. The **second** and **fifth** shapes are triangles. The **third** and **fourth** shapes are circles. The **sixth** shape is a square.

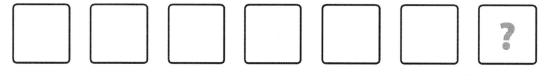

b What do you think the next shape is?

Continued overleaf

 Further mastery – number and place value **YEAR 1**

Name .... Class .... Date ....

4 Write the number **32** in the correct place on each grid below.

a
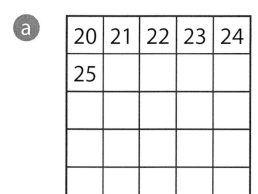

20	21	22	23	24
25				

b

20	21	22
23	24	25

5 Write the missing numbers in the boxes.

a **17** is **1** less than ⬜ .

b **24** is **1** more than ⬜ .

c **1** more than **39** is ⬜ .

d ⬜ is **1** less than **59**.

e **79** is **1** less than ⬜ .

6 Put the missing numbers in each section. Look carefully.

a

| 56 | | | | 51 | | | | 47 | | |

b

| 8 | 10 | | | | | 22 | | | | |

c

| 5 | | | 20 | 25 | | | | | | |

d

| | 90 | | | | | 40 | | | 10 | |

Name .... Class .... Date ....

Further mastery – addition and subtraction

1 Complete the pattern by shading rows **3**, **4** and **5**.
Rows **1** and **2** are completed.
Complete the shading and number sentences for rows **6**, **7** and **8**.

Row 1 ● ● ● ● ● ● ●   [0] + [7] = [7]

Row 2 ○ ● ● ● ● ● ●   [1] + [6] = [7]

Row 3 ○ ○ ○ ○ ○ ○ ○   [2] + [5] = [7]

Row 4 ○ ○ ○ ○ ○ ○ ○   [3] + [4] = [7]

Row 5 ○ ○ ○ ○ ○ ○ ○   [4] + [3] = [7]

Row 6 ○ ○ ○ ○ ○ ○ ○   [ ] + [ ] = [ ]

Row 7 ○ ○ ○ ○ ○ ○ ○   [ ] + [ ] = [ ]

Row 8 ○ ○ ○ ○ ○ ○ ○   [ ] + [ ] = [ ]

2 The number sentences can be made from the pieces of fruit shown.

a Complete the missing numbers.

[4] + [2] = [6]
[ ] + [4] = [6]
[6] – [2] = [ ]
[6] – [ ] = [2]

b Now write **4** number sentences for the pictures below.

[ ] + [ ] = [ ]
[ ] + [ ] = [ ]
[ ] – [ ] = [ ]
[ ] – [ ] = [ ]

Continued overleaf

91

Further mastery – addition
and subtraction   **YEAR 1**

3 Complete the first number sentence and then use it to write **3** more sentences. An example is shown.

| 3 | + | 2 | = | 5 |

| 2 | + | 3 | = | 5 |

| 5 | − | 2 | = | 3 |

| 5 | − | 3 | = | 2 |

a [7] + [3] = [ ]

[ ] + [ ] = [ ]

[ ] − [ ] = [ ]

[ ] − [ ] = [ ]

b 15 − 8 = [ ]

[ ] − [ ] = [ ]

[ ] + [ ] = [ ]

[ ] + [ ] = [ ]

---

4 **a** Complete the following.

4 + [ ] = 4   5 + [ ] = 6   7 − [ ] = 7

8 − [ ] = 2   12 − [ ] = 10   13 + [ ] = 13

b Use your answers to complete the sentences below.

If you subtract **0** from a number, the answer ....

If you add **0** to a number, the answer ....

---

5 Roha has **4** fewer toy cars than Leo. Leo has **13** toy cars.
Write a number sentence to work out how many toy cars Roha has.

[ ] O [ ] = [ ]

6 Make up a problem to match this number sentence. 5 + 3 = 8

....

....

Name .... Class .... Date ....

Further mastery – multiplication and division

1 Megan is counting in **tens**, starting with **20**. Maisie is counting in **fives**, starting with **15**. Maisie says, "If we carry on counting, we will both say the number **95**."

Is Maisie correct? ☐ Try to explain your answer.

....

2 There are **6** trays of biscuits. Each tray has the same number of biscuits.

How many biscuits are there altogether? ☐

3 Zak has **12** pencils. He puts **2** pencils in each pencil pot.

PENCILS

How many pencil pots does he fill? ☐

4 Zoya has **8** pencils. She puts **half** of the pencils on the desk.

How many pencils does Zoya put on the desk? ☐

Continued overleaf

5 Molly has **16p**. Put a tick (✓) by the purses which could be Molly's.

a

☐

c

☐

b

☐

d

☐

6 Now draw coins in these purses to equal **12p** in each purse. Make each purse different.

7 Orla has **3** strawberries.

a Jamal has **double** the number of strawberries Orla has.

How many strawberries does Jamal have? ☐

b May has **5 times** as many strawberries as Jamal.

How many strawberries does May have? ☐

c Aftab has **one third** as many strawberries as May.

How many strawberries does Aftab have? ☐

Name .... Class .... Date........................

Further mastery – fractions

1 Shade **half** of each shape.

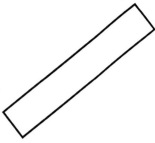 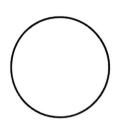 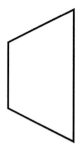

2 Tick (✓) the shapes which have **one quarter** shaded.

 ☐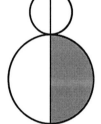

 ☐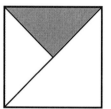

☐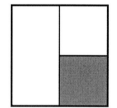

☐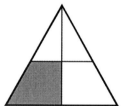

3 Complete the shading so that a **half** is <u>not</u> shaded.

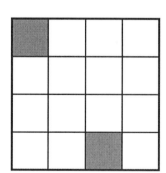

4 Draw rings so that a **quarter** of the chocolates are in each ring.

Continued overleaf

5 Write the correct numbers in each box.

20			

6 There are **18** children in a swimming pool.
Half of the children are boys.
10 of the children are girls.

Can this be correct? [ ]

Explain your answer.

....

....

7 Hugo, Max, Sean and Tegan share each type of fruit equally between them.

a Complete the table to show how many of each fruit Max gets.

strawberries	apples	bananas

b What fraction of the fruit does Tegan get? [ ]

c What fraction of the fruit do Hugo and Sean get together? [ ]

Further mastery – measurement

1 Which path is longer?

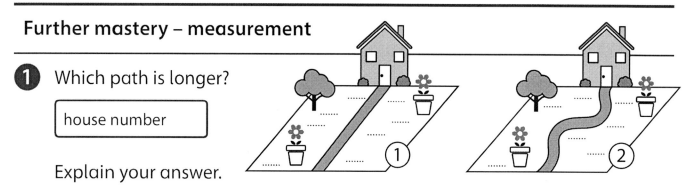

house number

Explain your answer.

....

2 Look at the scales below.

a Which toy is the heaviest? Circle your answer.

b Which toy is the lightest? Circle your answer.

3 Complete the **second** scales below to show what you think would happen, if the doll and the car were together on **one** side of the scales, and the drum was on the other.

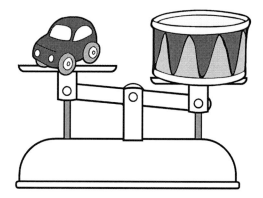

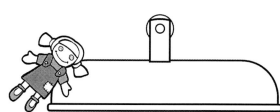

Continued overleaf

4 Rose had **two** glasses full of juice. She poured them into the jug. Circle the container with the greater capacity.

5 The **first** clock shows the time Alfie sets off for school. Lucy sets off **half an hour** later than Alfie and Stefan sets off **half an hour** later than Lucy. Complete the clocks below.

time Alfie
sets off

time Lucy
sets off

time Stefan
sets off

6 Put **T** (true) or **F** (false) by each of the statements.

a) **1 year** is shorter than **1 week**.

b) **1 day** is shorter than **1 week**.

c) **1 month** is shorter than **1 week**.

d) **1 minute** is shorter than **1 week**.

e) **1 hour** is shorter than **1 minute**.

Further mastery – geometry

1 Draw arrows to put each item in a circle. Write the name of the shapes in each circle. An example is shown.

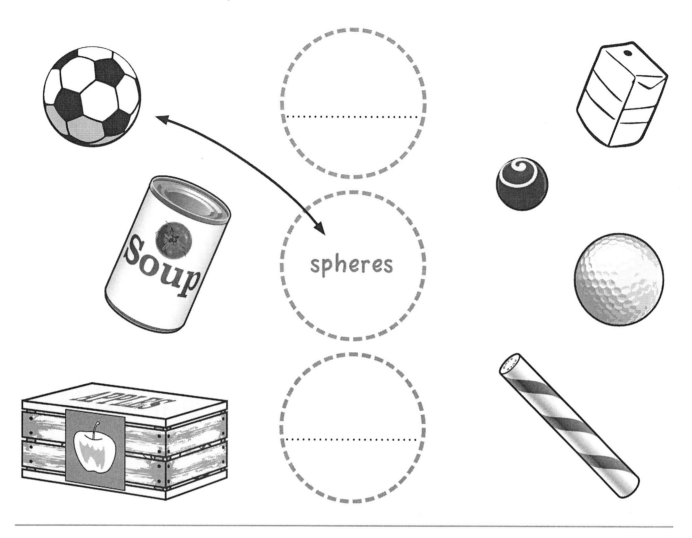

2 Put a tick by the odd one out.

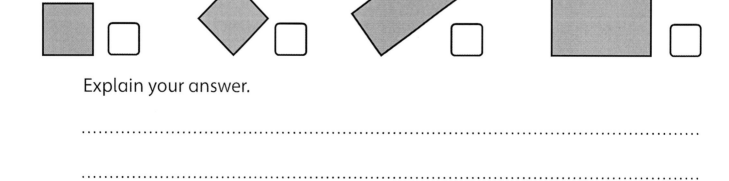

Explain your answer.

....

....

Continued overleaf

Name .... Class .... Date ....

a Draw a circle ◯ in the top row, **third** from the right.

b Draw a triangle △ in the bottom row, **second** from the left.

c Describe the position of the bird.

....

d Put a cross ✗ underneath the bird.

e Put a dot ● next to the circle, on top of the bird.

f Put **2** dots ● ● in the top right hand corner.

Further mastery – geometry   **YEAR 1**

Answers

ANSWERS

Page 1

1 Children counting correctly

2 Children counting correctly

3 a) 3, 4, 5 b) 17, 18, 19 c) 28, 29, 30
   d) 98, 99, 100 e) 101, 102, 103

Page 2

1 Children counting correctly

2 Children counting correctly

3 a) 6, 5 b) 16, 15 c) 25, 24, 23
   d) 82, 81, 80 e) 100, 99, 98

Page 3

1 matched correctly

2 a) third b) fifth

3 4th, 6th, 7th, 8th, 10th, 11th

Page 4

1 6

2 15

3 38

4 20 cm

5 62

Page 5

1 a) 8 b) 12 c) 16

2 a) 6, 8, 10, 12 b) 12, 16, 18, 20 c) 30, 32, 36, 38
   d) 21, 23, 25, 27

Page 6

1 a) 15 b) 20 c) 35

2 a) 10, 15, 20, 25 b) 35, 40, 45, 50
   c) 70, 75, 80, 85 d) 38, 43, 48, 53

Page 7

1 a) 40 b) 60 c) 90

2 a) 20, 30, 40, 50 b) 60, 70, 80, 90 c) 75, 95, 105,
   115 d) 43, 63, 73, 83

Page 8

1 rings around 17, 3, 9

2 2, 4, 6, 8, 10, 12, 14, 16, 18

3 21, 23, 25, 27, 29, 31, 33, 35, 37, 39

4 10, 20, 30, 40, 50, 60, 70, 80, 90

5 a) 2nd column: 77, 83, 89, 95; 4th column: 79, 85,
   91, 97; 6th column: 81, 87, 93, 99 b) appropriate
   explanation, e.g. odd numbers in columns,
   increasing in 6 because 6 numbers in a row

Page 9

1 a) 9 b) 11 c) 16

2 a) 8 b) 10 c) 37 d) 80 e) 101 f) 16 g) 99

Page 10

1 a) 6 b) 11 c) 23

2 a) 5 b) 20 c) 13 d) 37 e) 99 f) 62 g) 81

Page 11

1 1, 2, 3

2 11, 13, 14

3 a) 18, 19, 20, 21 b) 39, 40, 42, 43 c) 93, 95, 97, 99

4 row 1: 12, 13, 14, 15, 16, 17, 18;
   row 2: 21, 22, 23, 25, 26, 27;
   row 3: 34, 35, 36; Row 4: 46

Page 12

1 a) fewer b) more c) fewer d) equal

2 a) more than b) equal to c) less than
   d) more than e) equal to

Page 13

1 a) doll b) car

2 a) □ □ ○ b) □ △ ○

3 a) ✈, 🚲 b) ○ □ c) ○ ○

Page 14

1 15: fifteen, 12: twelve, 4: four, 20: twenty, 10: ten

2 a) 3 b) 2 c) 8 d) 17 e) 11 f) 19

3 a) seven b) eleven c) fourteen d) sixteen
   e) twenty

Page 15

1 27

2 19

3 a) 20, 25, 30, 35, 40, 45 b) 20, 30, 40 circled

4 52, 50, 48, 46, 44, 42

Page 16

5) 17, seventeen

6) a) even b) odd c) even d) odd

7) 30

8) 30

Page 17

1 – : subtract, + : add, = equals

2 a) + b) +, = c) –, =

3 a) + b) – c) = d) = e) + f) –

ANSWERS

Page 18
1 a) 13 b) 13 c) 8 d) 8
2 a) 10 b) 6 c) 7 d) 7
3 a) 5 b) 8 c) 4 d) 3

Page 19
1 a) 6 dots drawn b) 4 dots drawn
   c) 5 dots drawn d) 5 dots drawn
2 a) 12 b) 5 c) 0 d) 7
3 a) 0 + 10, 1+ 9, 2 + 8, 3 + 7, 4 + 6, 5 + 5 b) 4, 7, 2, 0

Page 20
1 0 + 20, 1 + 19, 2 + 18, 3 + 17, 4 + 16, 5 + 15,
   6 + 14, 7 + 13, 8 + 12, 9 + 11, 10 + 10
2 a) 13 + 0, 9 + 4, 2 + 11, 3 + 10, 12 + 1, 9 + 4
   b) 3, 13, 4, 12
3 a) 4, 4 b) 0, 0 c) 15, 15

Page 21
1 a) 7 b) 2 c) 10 d) 12
2 a) 8 b) 4 c) 6 d) 11 e) 13 f) 15
3 a) 9 b) 18 c) 15 d) 6 e) 17 f) 16: number

Page 22
1 a) 14 b) 14 c) 20 d) 17 e) 18 f) 18 g) 17 h) 17
2 a) ring around: 11 + 8, 17 + 2, 14 + 5, 13 + 6
   b) ring around: 17 + 0, 11 + 6, 14 + 3
   c) ring around: 12 + 6, 13 + 5,
   d) ring around: 1 + 13, 14 + 0, 11 + 3
   e) ring around: 11 + 4, 10 + 5, 13 + 2

Page 23
1 5
2 a) 6 b) 4
3 a) 3 b) 6 c) 7 d) 0
4 a) 3 b) 6 c) 6 d) 7 e) 4 f) 0 g) 4 h) 5

Page 24
1 a) 12 b) 12 c) 7 d) 4 e) 8 f) 7 g) 12 h) 0 i) 4
   j) 13
2 a) 18 – 5 = 13, 19 – 4 = 15, 12 – 8 = 4 (shaded),
   17 – 13 = 4 (shaded), 19 – 11 = 8 (shaded),
   b) 18 – 12 = 6, 14 – 5 = 9 (shaded),
   17 – 17 = 0, 19 – 1 = 18 (shaded),
   15 – 6 = 9 (shaded)

Page 25
1 a) 1 b) 13 c) 8 d) 9 e) 11 f) 0 g) 17 h) 9
   i) 6 j) 15
2 a) 6, 15, 3 b) 1, 20, 8, 9
   c) appropriate calculations equalling 11

Page 26
1 9
2 10
3 18p
4 20 cm

Page 27
5 15
6 17
7 17p
8 13

Page 28
1 2
2 5
3 6
4 12 cm

Page 29
5 7
6 5
7 9p
8 9p

Page 30
1 17
2 14
3 9 m
4 18p

Page 31
5 4
6 27p
7 a) 8 m b) 17 m
8 12

ANSWERS

Page 32
1   2, 4
2   2, 8
3   3, 15
4   4, 40

Page 33
5   3, 2, 6
6   3, 5, 15
7   2, 10, 20
8   6, 2, 12

Page 34
1   8
2   30
3   30
4   6

Page 35
5   15
6   16
7   90
8   60

Page 36
1   5, 15
2   4, 40
3   5, 20
4   2, 16

Page 37
5   3, 10, 30
6   2, 7, 14 or 7, 2, 14
7   4, 10, 40 or 10, 4, 40
8   3, 5, 15; 5, 3, 15

Page 38
1   6, 2, 12
2   5, 4, 20
3   3, 10, 30
4   15

Page 39
5   20
6   35
7   18
8   30

Page 40
1   20
2   14
3   30

Page 41
1   12
2   50
3   15
4   40

Page 42
5   14p
6   40
7   24
8   30 eggs, 10 glasses of milk, 40 spoonfuls of sugar, 20 cups of flour

Page 43
1   carrots drawn in each box: a) 4 b) 2 c) 8
2   1st column top to bottom: 6, 0, 10;
     2nd column top to bottom: 12, 16, 20, 40

Page 44
1   6
2   8
3   18p
4   24
5   40 centimetres

Page 45
1   2, 5
2   2, 6
3   10, 2
4   3, 3

Page 46
5   10, 5, 2
6   15, 3, 5
7   20, 5, 4
8   18, 6, 3

ANSWERS

Page 47

1   4
2   3
3   3
4   8 dots in each circle

Page 48

5   3
6   3
7   4 metres
8   6

Page 49

1   3
2   8
3   3
4   7

Page 50

5   3
6   4p
7   8
8   5

Page 51

1   5, 3
2   5, 2
3   15, 2
4   4, 5

Page 52

5   3, 4; 4, 3
6   4, 6; 6, 4
7   7, 2; 2, 7
8   20, 5, 4; 20, 4, 5

Page 53

1   3
2   appropriate array: 4
3   appropriate array: 3
4   appropriate array: 4

Page 54

1   5
2   4
3   7
4   7

Page 55

5   6
6   6p
7   6 metres
8   8p

Page 56

1   ring around: a) 3 sheep b) 4 pigs
2   ring around: a) 1 square b) 2 triangles
3   a) 2 b) 5 c) 3 d) 8 e) 5

Page 57

1   3
2   5
3   10p
4   2

Page 58

5   a) 8 b) 4
6   5
7   a) 12 centimetres b) 6 centimetres
     c) 8 centimetres

Page 59

1   10
2   30
3   3
4   a) 2 kilograms b) 16 kilograms

Page 60

5   14 years old
6   5
7   24
8   a) 6p b) 60p c) 15p d) 5

Page 61

1   a) half b) half
2   a) half b) equal to circled
3   half shaded correctly
4   the tie ticked

ANSWERS

Page 62

1 a) half b) half
2 half of each shape shaded correctly
3 shapes a, c and d ticked
4 one or one whole

Page 63

1 half
2 half
3 7 buttons shaded
4 a) 1 b) 4 c) 6 d) 8 e) 11

Page 64

1 a) half b) half
2 arrow at 5 cm
3 a) giraffe b) cow, dog

Page 65

1 a) equals b) quarter
2 a) completed correctly b) quarter c) 4

Page 66

1 quarter
2 lines drawn correctly
3 ticks by c and d

Page 67

1 quarter
2 quarter
3 ring around 3 children
4 a) 4 b) no: appropriate to explain they both have one quarter

Page 68

1 a) arrow at 1 km b) quarter
2 lines drawn at 4 cm and 6 cm
3 quarter

Page 69

1 2
2 4 candles drawn
3 2 metres
4 a) quarter b) £3

Page 70

5 a) 4 b) 4 c) yes: appropriate to explain 2 quarters equals one half
6 a) 3 b) 6 c) yes: appropriate to explain one half equals 2 quarters d) quarter

Page 71

1 tick by the pencil
2 tick by the lamp post
3 a) shorter b) taller, higher or longer
4 3

Page 72

1 tick by the horse
2 circle around the elephant, tick by the fish
3 a) lighter than b) heavier than

Page 73

1 tick by 2nd jug
2 lines drawn correctly
3 a) less b) more
4 container

Page 74

1 tick by 'pouring a glass of water'
2 tick by 'walking home from school'
3 a) quicker b) before
4 Kira

Page 75

1 9 units
2 6 units
3 10 centimetres
4 a) 4 centimetres b) 7 centimetres c) 8 centimetres

Page 76

1 ticks by a and d
2 a) 4 kilograms b) 3 kilograms
3 a) 24 kilograms b) 55 kilograms

Page 77

1 ticks by a, b, c and f
2 from top: 1 litre, 600 millilitres, 300 millilitres, 100 millilitres

ANSWERS

Page 78

1 12 seconds circled

2 10 hours circled

3 50 minutes circled

4 a) reasonable amount of time b) reasonable amount of time

d 1 hour or 60 minutes d) 3 hours

Page 79

1 money and values correctly matched

2 ticks by a, b and d

3 2 different combinations, e.g 12 × 1p, 6 × 2p, 10p + 2p, 10p + (1p × 2), (5p × 2) + 2p, (5p × 2) + (1p × 2)

Page 80

1 first, next, before, later, morning, evening, after, until

2 tomorrow, today, yesterday

Page 81

1 Tuesday, Wednesday, Thursday, Friday, Saturday, Sunday

2 a) weeks b) days c) months

3 a) January b) April c) September d) December

4 a) 24 b) 2015 c) October

Page 82

1 a) 7 o'clock b) half past 2

2 a) 3 o'clock b) half past 4 c) half past 9
   d) 11 o'clock e) 12 o'clock f) half past 6

Page 83

1 times drawn correctly on the clock faces

2 times drawn correctly on the clock faces

3 a) 6 o'clock drawn correctly on the clock face
   b) 6:30 drawn correctly on the clock face

Page 84

1 tick by all on the top row and bottom right

2 triangle: appropriate explanation – 3 sides and 3 corners or vertices

3 appropriate drawing

4 No: appropriate explanation shape properties – 4 equal sides & 4 equal square (or 90°) angles

Page 85

1 Lines drawn correctly

2 a) sphere b) cube c) cone d) cuboid

3 No: appropriate explanation shape properties – cube has 6 equivalent square faces, cuboid can have oblong faces

Page 86

1 a) 6 b) 3 c) 3

2 a) trifle b) biscuit c) appropriate: makes half a turn in an anti-clockwise direction d) trifle

Page 87

1 a) lorry b) ice-cream van c) car d) top left

2 a) on top of b) underneath c) between
   d) inside e) around f) near

Page 88

3 a) forwards b) backwards c) close d) far
   e) up f) down

Page 89

1 a) tick by second road b) 11: eleven

2 pattern drawn correctly

3 a) □ △ ○ ○ △ □ b) triangle

Page 90

4 32 written: a) 3rd row, 3rd column
   b) 5th row, 1st column

5 a) 18 b) 23 c) 40 d) 58 e) 80

6 a) 55, 54, 53, 52, 50, 49, 48, 46, 45
   b) 12, 14, 16, 18, 20, 24, 26, 28, 30
   c) 10, 15, 30, 35, 40, 45, 50, 55, 60
   d) 100, 80, 70, 60, 50, 30, 20, 0

Page 91

1 row 3: 2 dots not shaded, 5 shaded;
   row 4: 3 dots not shaded, 4 shaded;
   row 5: 4 dots not shaded, 3 shaded;
   row 6: 5 dots not shaded, 2 shaded, 5 + 2 = 7;
   row 7: 6 dots not shaded, 1 shaded, 6 + 1 = 7;
   row 8: 7 dots not shaded, 7 + 0 = 7

2 a) 2 + 4 = 6, 6 − 2 = 4, 6 − 4 = 2
   b) 5 + 4 = 9, 4 + 5 = 9, 9 − 5 = 4, 9 − 4 = 5

ANSWERS

Page 92

3   a) 10, 3 + 7 = 10, 10 − 7 = 3, 10 − 3 = 7
    b) 7, 15 − 7 = 8, 7 + 8 = 15, 8 + 7 = 15

4   a) 4 + 0 = 4, 8 − 6 = 2, 5 + 1 = 6, 12 − 2 = 10,
    7 − 0 = 7, 13 + 0 = 13
    b) appropriate to explain when adding &
    subtracting 0 to and from a number, the number
    remains the same

5) 13 − 4 = 9

6) appropriate, e.g. There are 5 girls & 3 boys.
    How many children are there altogether.

Page 93

1   No: appropriate explanation that 95 is not a
    multiple of 10

2   60

3   6

4   4

Page 94

5   Tick by a, c and d

6   2 different combinations drawn, e.g. 10p + 2p,
    5 × 2 + 2p, 5p × 2 + 1p × 2, 1p × 12, 2p × 6

7   a) 6 b) 30 c) 10

Page 95

1   half of each shape shaded appropriately

2   tick by b and c

3   6 additional squares shaded

4   4 rings with 3 chocolates in each ring

Page 96

5   5 in each box

6   No: appropriate to explain that half of the
    children must be girls and half of 18 is 9

7   a) 2 strawberries, 1 apple, 3 bananas b) quarter
    c) half

Page 97

1   2: appropriate, to explain length of both gardens
    is equal, path 2 is wavy and therefore longer

2   a) drum circled b) doll circled

3   scale drawn to show the left hand pan is lower
    than on the first scale

Page 98

4   jug circled

5   8 o'clock drawn on 2nd clock,
    8:30 drawn on 3rd clock

6   a) F b) T c) F d) T e) F

Page 99

1   cylinders: arrows from soup and rock;
    spheres: arrows from chocolate and golf ball
    cuboids: arrows from crate and drink carton

2   tick by the oblong: explanation that the rest are
    squares/ have equal sides & square corners

Page 100

3   a) & b) placed appropriately
    c) appropriate, e.g. middle row, 3rd from left
    d), e) & f) placed appropriately